Palgrave Studies in the History of Science and Technology

Designed to bridge the gap between the history of science and the history of technology, this series publishes the best new work by promising and accomplished authors in both areas. In particular, it offers historical perspectives on issues of current and ongoing concern, provides international and global perspectives on scientific issues, and encourages productive communication between historians and practicing scientists.

More information about this series at
http://www.palgrave.com/gp/series/14581

Katharina C. Cramer

A Political History of Big Science

The Other Europe

palgrave
macmillan

Katharina C. Cramer
University of Konstanz
Konstanz, Germany

Palgrave Studies in the History of Science and Technology
ISBN 978-3-030-50048-1 ISBN 978-3-030-50049-8 (eBook)
https://doi.org/10.1007/978-3-030-50049-8

This Palgrave Macmillan imprint is published by the registered company Springer Nature Switzerland AG.
The registered company address is: Gewerbestrasse 11, 6330 Cham, Switzerland

This book is dedicated to Justus and Cornelius, who have grown alongside this project.

Preface

This book is about light. It explores the history and science of the brilliant light, synchrotron radiation, that is produced at two collaborative light sources in Europe, namely, the ESRF as a circular-shaped synchrotron radiation source and the European XFEL as a linear free-electron laser. In the early decades after the first experimental observation of synchrotron radiation in the late 1940s, research with synchrotron radiation was a marginal phenomenon in the scientific landscapes in Europe and the United States that were largely dominated by particle physics research. Probably nobody would have guessed at that time that synchrotron radiation would become one of the most crucial experimental resources for multidisciplinary research in the twenty-first century and a kind of mainstream activity for the investigation of materials or living matter. But this book also sheds new light on the history and politics of Big Science, Europe and the European Union. One of its core aims is to enlighten the ways we see, write and think about Europe and the European Union, as well as about European politics and history. It introduces *the other Europe* as an alternative perspective to politics and integration in Europe besides the mainstream political integration processes, arguing that Big Science collaborations, such as the ESRF and the European XFEL, have played crucial roles in both European politics and science.

This book is based on a doctoral dissertation that was carried out between 2014 and 2018 at the Leibniz Prize Research Group "Global Processes" at the University of Konstanz, Germany (date of oral examination: 30 August 2018, examiners: Jürgen Osterhammel, Olof Hallonsten and Anne Kwaschik). It is a great pleasure to thank my supervisors Jürgen Osterhammel and Olof Hallonsten for advice and support.

The book relies to a great extent on personal encounters, correspondences and interviews. The conduction of interviews served a very broad purpose, namely, to gain access to the larger community of scientists and administrators, to further identify key actors and close observers that played important roles during the establishment of the ESRF and the European XFEL and to get to know concerns that were missing or unlikely to ever be displayed in official documents. Only a very small part of the many interviews and correspondences were eventually used in this book. It is impossible to name all those who welcomed me with hospitality at DESY, ESRF and European XFEL, and who shared their knowledge and expertise. But I would like to thank explicitly Chantal Argoud, Cerstin Barmbrock, Karen Clugnet, Itziar Echeverría, Robert Feidenhans'l, Petra Folkerts, Nathalie Godet, Petra Hendrikman-Verstegen, Martin Köhler, Rainer Koepke, Olaf Kühnholz, Christof Kunz, Axel Lindner, Frieder Meyer-Krahmer, Denes Laos Nagy, Luis Sanchez Ortiz, Frank Poppe, Martin Sandhop, Hermann Schunck, Franscesco Sette, Christian Vettier, Renata Witsch, Karl Witte and Thomas Zoufal for their time and efforts.

I would also like to thank the members of the Leibniz Programme "Global Processes" at the University of Konstanz; the members of the Division of History of Science, Technology and Environment at KTH Stockholm; and Mats Benner, Thomas Kaiserfeld, Josephine Rekers and Maria Moskovko at Lund University for encouragement, comments and critics.

Alfter, Germany Katharina C. Cramer

Contents

Abbreviations

4GLS	4th Generation Light Source
ACO	Anneau de Collisions d'Orsay, Orsay Storage Ring
AEC	Atomic Energy Commission
AGF	Arbeitsgemeinschaft Großforschungseinrichtungen
ALICE	A Large Ion Collider Experiment
ALS	Advanced Light Source
ANKA	Angströmquelle Karlsruhe
APS	Advanced Photon Source
BER	Berlin Research Reactor
BESSY	Berliner Elektronenspeicherring-Gesellschaft für Synchrotronstrahlung mbH, Berlin Electron Storage Ring Society for Synchrotron Radiation
BMBF	Bundesministerium für Bildung und Forschung, Federal Ministry for Education and Research (since 1998)
BMFT	Bundesministerium für Forschung und Technologie, Federal Ministry for Research and Technology (1972–1994)
BRITE	Basic Research in Industrial Technologies
CAP	Common Agricultural Policy
CCLRC	Council for the Central Laboratory of the Research Councils
CDR	Conceptual Design Report
CEA	Commissariat à l'Énergie Atomique, Atomic Energy Commission
CENT	Centre National d'Études des Télécommunications
CERN	European Organization for Nuclear Research, originally: Conseil Européen pour la Recherche Nucléaire
CESR	Cornell Electron Storage Ring

CHESS	Cornell High-Energy Synchrotron Source
CNRS	Centre National de la Recherche Scientifique, National Center for Scientific Research
COST	Cooperation Européenne dans le Domaine de la Science et de la Technologie, European Cooperation in Science and Technology
CREMLIN	Connecting Russian and European Measures for Large-Scale Research Infrastructures
DCI	Dispositif de Collisions dans l'Igloo
DELTA	Dortmund Electron Accelerator
DESY	Deutsches Elektronen-Synchrotron, German Electron Synchrotron
DOE	Department of Energy
DORIS	Doppel-Ring-Speicher, Double-Ring Storage
ECMST	European Center for Marine Science and Technology
ECSC	European Coal and Steel Community
EEC	European Economic Community
EIB	European Investment Bank
ELDO	European Space Vehicle Launcher Development
ELSA	Elektronen-Stretcher Anlage
EMBL	European Molecular Biology Laboratory
EMBO	European Molecular Biology Organization
EMU	European Monetary Union
EPSRC	Engineering and Physical Sciences Research Council
ERA	European Research Area
ERIC	European Research Infrastructure Consortium
ERL	Energy Recovery Linac
ERP	European Recovery Program
ESF	European Science Foundation
ESFRI	European Strategy Forum on Research Infrastructures
ESO	European Southern Observatory
ESPRIT	European Strategic Program on Research in Information Technology
ESRF	European Synchrotron Radiation Facility
ESRO	European Space Research Organisation
ESRP	European Synchrotron Radiation Project
ESS	European Spallation Source
ETW	European Transonic Wind Tunnel
EU	European Union
Euratom	European Atomic Energy Community
EUREKA	European Research Coordination Agency

EWR	Erweiterter Wissenschaftlicher Rat, Extended Scientific Council
FAIR	International Accelerator Facility for Beams of Ions and Antiprotons
FEL	Free-Electron Laser
FERMI	Free Electron Laser Radiation for Multidisciplinary Investigations
FLASH	Free Electron Laser in Hamburg
FP	Framework Programme for Research and Technological Development
GDP	Gross Domestic Product
GeV	Gigaelectron Volt
GmbH	Gesellschaft mit beschränkter Haftung
GSI	Gesellschaft für Schwerionenforschung, Society for Heavy Ion Research
HALO	High Altitude and Long Range Research Aircraft
HASYLAB	Hamburger Synchrotronstrahlungslabor, Hamburg Synchrotron Radiation Laboratory
HDL	High Field Laboratory Dresden
HERA	Hadron-Elektron-Ring-Anlage
HFBR	High Flux Beam Reactor
HLD	High Field Laboratory Dresden
HMI	Hahn Meitner Institute
ICFA	International Committee for Future Accelerators
IHEP	Institute for High Energy Physics
IKRC	In-Kind Review Committee
ILC	International Linear Collider
ILL	Institut Laue-Langevin
INSERM	Institut National de la Santé et de la Recherche Médicale
IR	infrared
IRAM	Institut de Radioastronomie Millimétrique
IRF	Institut de Recherche Fondamentale
IRI	Ioffe-Röntgen Institute
ITER	International Thermonuclear Experimental Reactor
JET	Joint European Torus
KEK	Japanese acronym for: High Energy Accelerator Research Organisation
KIT	Karlsruhe Institute of Technology
km	kilometre
LCLS	Linear Coherent Light Source
LEP	Large Electron Positron Collider
LHC	Large Hadron Collider
linac	linear accelerator
LOP	Loi d'Orientation et de Programmation pour la Recherche et le Développement Technologique de la France

LURE	Laboratoire pour l'Utilisation du Rayonnement Électromagnétique
LUSY	Lund University Synchrotron
m	metre
MAX	Microtron Accelerator for X-rays
MeV	megaelectron volts
MoU	Memorandum of Understanding
MPG	Max-Planck-Gesellschaft, Max Planck Society
MRC	Medical Research Council
MSR	Medium Flux Reactor
MST	Mission Scientifique et Technique, Mission on Science and Technology
NATO	North Atlantic Treaty Organization
NINA	Northern Institute for Nuclear Accelerators
NLS	New Light Source
nm	nanometre
NSLS	National Synchrotron Light Source
NTF	National Transonic Facility
OECD	Organisation for Economic Co-operation and Development
PAL XFEL	Pohang Accelerator Laboratory X-ray Free-Electron Laser
PEP	Positron-Electron Project
PETRA	Positron-Elektron Tandem Ring Anlage, Positron-Electron Tandem Ring Accelerator
PPARC	Particle Physics and Astronomy Research Council
PREST	Politique de Recherche Scientifique et Technologique
PSI	Paul Scherrer Institute
QuBS	Quantum Beam Science
RACE	Research and Development in Advanced Communications Technologies in Europe
RAMIRI	Realising and Managing International Research Infrastructures
ROSATOM	Rosatom State Atomic Energy Corporation
RUSNANO	Russian Corporation of Nanotechnologies
SACLA	SPring-8 Angstrom Compact Free Electron Laser
SASE	Self-Amplified Spontaneous Emission
SBLC	s-Band Linear Collider
SDUV-FEL	Shanghai Deep-Ultraviolet Free Electron Laser
SERC	Science and Engineering Research Council
SINAP	Shanghai Institute of Applied Physics
SLAC	Stanford Linear Accelerator Center
SLS	Swiss Light Source
SNQ	Spallations-Neutronenquelle
SNS	Spallation Neutron Source

SOHO	Solar and Heliospheric Observatory
SOLEIL	Source Optimisée de Lumière d'Énergie Intermédiaire du LURE
SPEAR	Stanford Positron Electron Asymmetric Rings
SPring-8	Super Photon Ring-8 GeV
SPS	Super Proton Synchrotron
SRF	Synchrotron Radiation Facility
SRS	Synchrotron Radiation Source
SSC	Superconducting Super Collider
SSRL	Stanford Synchrotron Radiation Lightsource Division
STFC	Science and Technology Facilities Council
SuperACO	see: ACO
SuperKEKB	KEK-B-factory, see: KEK
SXFEL	Shanghai Soft X-rays Free Electron Laser
TDR	Technical Design Report
TESLA	Tera-Electronvolt Energy Superconducting Linear Accelerator
TRISTAN	Transposable Ring Intersecting Storage Accelerator in Nippon
TTF FEL	Free-Electron Laser at the TESLA Test Facility
TTF	TESLA Test Facility
UV	ultraviolet
VAT	Value Added Tax
VEPP	Russian acronym for: Colliding Electron Beams
VUV	vacuum-ultraviolet
WR	Wissenschaftlicher Rat, Scientific Council
XFEL	X-ray Free Electron Laser
XUV	extreme ultraviolet

The following names of projects, accelerators and/or light sources that are used in this thesis do not constitute acronyms and/or abbreviations: Alba, Diamond, Elettra, ISIS, Tantalus and Aladdin.

List of Figures

List of Tables

CHAPTER 1

Introduction: History and Politics of Big Science in Europe

From single large instruments such as particle accelerators, telescopes, neutron reactors, synchrotron radiation sources or free-electron lasers, to networks, distributed research infrastructures or cloud-based efforts, Big Science projects have become crucial and vital elements of the European scientific landscapes since the second half of the twentieth century. These projects are precious but also crucial resources with regard to the importance of their performances for the advancement of science together with the observation that their efforts are hardly duplicated at any other place in Europe or elsewhere. The political expectations that are nowadays placed on publicly funded Big Science projects are high namely that they should considerably contribute to the solving of urgent societal challenges, such as climate change, health or energy security.[1]

Several collaborative and single-sited Big Science facilities with different scientific purposes were established in Europe over the course of the last decades. The creation of CERN (European Organization for Nuclear Research) in 1954, ESRO (European Space Research Organisation) in 1962, ELDO (European Space Vehicle Launcher Development) in 1964, ILL (Institut Laue-Langevin) in 1966, EMBL (European Molecular Biology Laboratory) in 1973, ESRF (European Synchrotron Radiation Facility) and ETW (European Transonic Wind Tunnel) both in 1988 and European XFEL (European X-ray Free-Electron Laser) in 2009 are only some of the many projects of this kind. Intergovernmental agreements by

K. C. Cramer, *A Political History of Big Science*,
Palgrave Studies in the History of Science and Technology,
https://doi.org/10.1007/978-3-030-50049-8_1

state groups of varying size, negotiated among ministerial and governmental representatives, have become, and remain, the widespread *modus operandi* of these Big Science projects in Europe. Based on loosely structured ad-hoc processes that preceded their establishment, every project became, for better or worse, a unique piece within the scientific and political landscapes of Europe.

This book investigates the political history of Big Science in Europe characterised by the founding histories of two collaborative, single-sited facilities, namely the ESRF (European Synchrotron Radiation Facility) in Grenoble, France and the European XFEL (X-ray Free-Electron Laser) in Schenefeld, Germany. The ESRF was (and remains) the first collaborative synchrotron radiation facility in Europe. It was established in 1988 through intergovernmental agreement among eleven European countries that were Belgium, Denmark, Finland, France, Germany,[2] Italy, Norway, Spain, Sweden, Switzerland and the United Kingdom. Based on recommendations from leading European scientists to set up a collaborative effort on research with synchrotron radiation, the project developed under the auspices of the ESF (European Science Foundation) and in the context of intergovernmental negotiations mainly between France, Germany and the United Kingdom. The convention was signed in 1988, and the ESRF became operational in 1994.

The European XFEL is a free-electron laser that operates in the hard X-ray wavelength regime. The project is based on intergovernmental agreement that was signed in 2009 by twelve countries: Denmark, France, Germany, Greece, Hungary, Italy, Poland, Russia, Slovakia, Spain, Sweden and Switzerland. The founding history of the European XFEL project is closely connected to the activities of the international TESLA (Tera-Electronvolt Energy Superconducting Linear Accelerator) collaboration located at the German national research centre DESY (Deutsches Elektronen-Synchrotron). In the early 1990s, the TESLA collaboration had proposed the construction of a linear collider for research in particle physics. For various reasons which are to be explored in the context of this book, a free-electron laser was added several years later to the initial project proposal. In 2003, the German government decided to realise the free-electron laser, but to put a halt to the linear collider project. While the linear collider project was hence abandoned, the convention for the free-electron laser project was signed in 2009. The facility opened to external users in 2017.

The ESRF and the European XFEL produce intense and brilliant light: synchrotron radiation. This is a specific kind of electromagnetic radiation that was first discovered in the late 1940s at a synchrotron, a circular-shaped particle accelerator, from which this name derives.[3] Synchrotron radiation became an increasingly demanded experimental resource for multidisciplinary investigations into materials and living matter, as well as the development of drugs or smart materials. Today, nearly all research with synchrotron radiation is done at storage rings (another kind of circular-shaped particle accelerator) and free-electron lasers, which are based on a linear accelerator complex (see Chap. 3). Nevertheless, the (misleading) notion of synchrotron radiation has stuck among scientists, administrators, as well as in the public mind, and is also used throughout this book. An alternative way of framing research with synchrotron radiation is to consider it as a part of the field of photon science, which is, very simply speaking, science with light. The ESRF and the European XFEL are so-called user facilities or service facilities that provide synchrotron radiation as an experimental resource to external users. The facilities are publicly funded, and access for fundamental, non-proprietary research groups to the ESRF and the European XFEL is granted on the basis of a scientific peer-review process. Both facilities also offer the possibility to buy experimental time by commercial companies and similar industry-related organisations to carry out proprietary research.

The main motivation of this book is to explore the founding histories of the ESRF and the European XFEL, and to understand how these two Big Science collaborations came into being in the late twentieth and early twenty-first centuries. What were the main motivations to initiate and join these two collaborative Big Science projects? How were national research policy strategies and scientific needs set and negotiated? How did one compromise on site, financial share and legal framework? These questions are fundamental not only to understand the history and politics of the ESRF and the European XFEL but also to gain a nuanced understanding of how their founding histories relate and connect to the broader patterns and dynamics of European politics, European integration and international relations.

More than three decades after the convention of the ESRF was signed in 1988, and more than one decade after the signing of the convention of the European XFEL in 2009, the political processes that preceded both events remain largely unexplored events in the history of science and technology and the history of Europe.[4] Based on largely unexplored material from the French national archives (Archives Nationales de France), the German national archive (Bundesarchiv) and the internal archives of DESY and the ESRF, as well as through the analysis of specific scientific and political case-related dynamics, this book hopes to contribute to an improved understanding of the history and politics of Big Science in Europe.

This book partakes in a generational shift that is currently taking place in the study of Big Science. Current research efforts have started to broaden the disciplinary angles of the study of Big Science (such as political science[5] or innovation studies[6]) and to explore various new thematic fields (such as research infrastructures for the humanities[7] or evolving EU policy around Big Science projects[8]). But they also expand the (historical) study of Big Science well into the twenty-first century.[9] Scholarly research began to frame a narrative of change and continuity in the politics and organisation of Big Science projects in Europe, arguing that politics, economy, scientific programmes and organisation of Big Science profoundly changed throughout the late twentieth century and the early twenty-first century, while key principles and basic infrastructures largely remained in place.[10] Such a perspective does not only highlight how the history of Big Science considerably refrains and mirrors the historical development of European politics and policymaking.[11] But it also points to changes in the science policy rationales, most notably in the post-Cold War, attributing a more strategic role to knowledge, science and research for and within economy and society.[12] This also translated into a re-direction of funding priorities and rationales for the support of and commitment to Big Science. Such and similar emerging perspectives on the politics and organisation of Big Science have attracted considerable interest in recent years, most notably under the notions of *Big Science Transformed*[13] or *New Big Science*.[14] These approaches share two common denominators: First, they question traditional understandings of Big Science as a Cold War phenomenon.[15] But they further relate them to contemporary developments such as the emergence of Research Infrastructures (RIs) and the formation of a common[16] RI policy in Europe in the recent two decades (see below).[17] Second, they put emphasis on the investigation of the history and politics of synchrotron radiation sources, free-electron lasers and neutron sources since the late twentieth century. Organisation and framework of these scientific fields and experimental resources stand in considerable contrast to large particle physics projects that dominated Cold War Big Science and the study hereof.[18] With regard to research at synchrotron radiation sources, free-electron lasers and neutron sources, current research characterises its historical development since the second half of the twentieth century as gradual and stepwise that had started on a small scale. It remains small-scale when compared to major investments that still go into particle physics facilities such as CERN, although the discipline experienced decline throughout the last decades with regard to its importance, prestige and unprecedented growth rates throughout the Cold War.[19]

The notion of Big Science[20] has probably become the most prominent way to address scientific projects that are particularly large in terms of size, funding, manpower, organisational framework or political relevance and expectations. Although Big Science serves as an attractive buzzword to gather scholarly, public and political interest, in most cases, it remains an elusive concept. Scholarly research has long struggled to properly define and frame Big Science and remains to do so.[21] Historians of science James Capshew and Karen Rader proposed to differentiate between *big science* and *Big Science*; the latter one in capital letters "as a rhetorical construction,"[22] pointing to the particular dynamics of large-scale research following the end of the Second World War. Science administrator Pierre Papon similarly argued that efforts in establishing and funding Big Science projects in Europe in the post-war and Cold War context "opened a new era for European science."[23] Former director of the US-American Oak Ridge National Laboratory Alvin Weinberg and historian of science Derek De Solla Price, who were among the first to use the term Big Science in the 1960s, considered it as a particular condition of modern science. For Weinberg, who worried about the consequences of Big Science becoming too big, it was a "pathological condition."[24] For Price, it was the result of a historical development and an evolutionary process with an exponential growth curve that would, however, at a certain point in time level off.[25]

The aspect of physical size dominates many writings on Big Science. Most notably, because it refers to the size of scientific instruments that often provide the (material) baseline from which further concerns can be investigated, such as the organisation of large scientific projects in an industrial manner, or the hierarchical structure of large teams that are formed around large instruments.[26] With regard to Big Science during the Cold War, the aspect of size often accounted for the ever-increasing circumferences of circular-shaped particle accelerators in high-energy physics/particle physics[27] research, which were needed to reach ever-higher energies and to study ever-smaller constituents of matter. In other words, the increasing size of particle physics accelerators in the Cold War equalled increasing performances.[28] This logic became questionable in recent decades, not least with the cancellation of the US-American SSC (Superconducting Supercollider) project in the early 1990s. The SSC project originated in the early 1980s as a major US-American effort in particle physics which, by its size, costs and complexity, would have easily outperformed any other effort in particle physics research at this time, most notably activities at CERN in Europe. The eventual cancellation of the project had many different reasons, but most importantly, it constituted the first time that particle physicists did not get their next (larger) accelerator funded (see also Sect. 5.2).[29]

Moreover, exclusive focus on physical size also risks to miss other kinds of Big Science beyond the disciplines of particle physics, most importantly with regard to the growing scholarly interest in the study of synchrotron radiation sources, free-electron lasers and neutron sources. These instruments and machines were not necessarily bigger and/or larger than their predecessors but often more powerful, more complex, faster or brighter. While most particle physics accelerators were designed for the discovery of a specific particle, force and/or interaction, the design of synchrotron radiation sources and free-electron lasers is rather open-ended and multipurpose (see Chap. 3). This means that these facilities, as service facilities, accommodate scientists from a broad variety of disciplines on a short-term basis, to which they provide brilliant light and experimental opportunities to study and investigate samples and materials.[30]

But are there good reasons to continue to use the rather traditional notion of Big Science despite the apparent need to challenge the analytically useless focus on physical size? Yes, because the bigness of Big Science does not necessarily hinge on physical size alone. But it also includes other perspectives that range, among others, from the scale of political controversy and/or conflict around the establishment of new Big Science projects to the degree of visibility in public discourse (which is probably much larger for CERN than for any other facility in Europe). Most importantly perhaps, while Big Science collaborations are built for science, they require political support and commitment to be funded and realised. Sociologist Olof Hallonsten defined Big Science along three dimensions, namely *big organisations*, *big machines* and *big politics*. Similarly, James Capshew and Karen Rader argued that "[f]ew could deny that Big Science was inherently political, since the accumulation of the necessary resources required the exercise of power."[31] Politics and policy apparently play a decisive role in collaborative, intergovernmental Big Science efforts when disparate national research priorities, financial shares, long-term commitments and site selection are negotiated in multilateral contexts that often lead to the conclusion of wider political package deals. Previous research also illustrated that the histories of Big Science collaborations resonate patterns and dynamics of bilateral and multilateral alliance-building. They also represent a way of containing the power of the other partner, framing diplomatic and political relationships, defining European space and territory, as well as the pursuit of national interests and strategies.[32] Most importantly, emphasis on big politics that surround the creation, construction and operation of Big Science projects does not deny size: Synchrotron radiation sources or free-electron lasers, around which this book centres, are indeed big in a physical sense. The resources, employees or infrastructures that are clustered around them are assembled on a much larger scale than is the case for smaller university-based projects or the like. Taken together, these observations certainly call for the continued use of Big Science.

To summarise, this book is less interested in a general perspective on the growth and spread of science and research activities or the physical size of single large scientific instruments. But it is keen to explore and analyse the big politics of Big Science, and, more precisely, the political history of Big Science in Europe uncovering processes of lobbying, negotiating, decision-making and institution-building. It would, however, be a naiveté to characterise Big Science as entirely politicised. The successful creation and implementation of several projects have to a considerable extent only been made possible through fundamental advancements in science and technology overcoming serious constraints that could otherwise have meant the end of the project. It should moreover be highlighted that the history of Big Science has also been driven forward by individuals that worked on new projects through tight formal decision-making processes and difficult political environments by tirelessly lobbying and promoting their scientific expertise and vision. In this regard, this book thus pays attention to both scientific *and* political contexts that open for systematic investigation and understanding of the patterns and dynamics in the recent history of Big Science in Europe (see Chaps. 2 and 3).

The politics and organisation of Big Science projects in Europe since the second half of the twentieth century are marked by parallel efforts, including both the pursuit of national agendas and the possibility to establish collaborative projects on an ad-hoc basis. In contrast, when Japan, the Soviet Union and the United States, which became large players in postwar and Cold War science, set up similar, competitive Big Science projects in the second half of the twentieth century, these were, first and foremost, established as national projects, such as KEK (High Energy Accelerator Research Organisation) in Japan or SLAC (Stanford Linear Accelerator Center) in the United States.[33] Intergovernmental scientific collaboration in Big Science was rare, although not completely absent, for these three countries.[34] In other words, to the extent that Big Science in Europe since the end of the Second World War was considerably marked by collaborative efforts, Big Science in the United States, Japan and the former Soviet Union/Russia remained dominated by solo efforts.

Since the second half of the twentieth century, many Big Science projects in Europe came into existence as a result of multilateral negotiations and collaboration based on intergovernmental agreement among a varied number of countries.[35] They are formally independent in the sense of being neither a body or institution of the EEC/EU nor confined to common policymaking. This was mainly due to a lack of a common research policy agenda, which only slowly changed at the end of the 1980s. Since the early 2000s, the EU began to expand its common competences in research policy. The European Commission did not only start to implement common measures it also initiated common policy agendas around Research Infrastructures (RIs). The European Commission introduced and began to use the concept of RIs in its policy documents since the early 2000s, which, however, lacks a clear and coherent definition.[36] There are good reasons to argue that the term RIs partly overlaps with that of Big Science because the European Commission also counts particularly large and complex instruments as well as user facilities among its RIs. But RIs, as defined by the European Commission, also encompass, for instance, data collections for the social sciences and humanities, computing grids or mobile air crafts.[37] These kinds of infrastructures neither are particularly big in a physical sense nor fit within a traditional understanding of Big Science (see above). But their founding histories probably also relate to big politics.

Summarising this current situation, the existence of collaborative Big Science projects in Europe in the twenty-first century is paralleled by increasing political expectations that the European Commission, as well as national European governments, put on the performances of RIs. However, the creation, construction and operation of Big Science projects in Europe remain to be based on intergovernmental agreement and thus formally disentangled from common EU policymaking, bodies and institutions. The history and politics of Europe and the EEC/EU[38] have nevertheless meant a lot for these intergovernmental Big Science projects to be established and realised. In other words, and this is one of the core messages of this book, patterns of diplomatic and political relations among countries in and around Europe, moments of deepening European integration, times of European political crisis and upheavals are nevertheless well resonated by the politics played out during the founding phases of Big Science collaborations.

The book pays attention to this particular situation in Europe through introducing the conceptual stance of *the other Europe* by exploring how Europe and the EEC/EU were framed, performed and established through the creation, construction and operation of Big Science collaborations in Europe. This perspective thus promotes a fresh look on the history and politics of Europe and the European integration process, and challenges the widely foregrounded research focus of European studies on treaty reforms and amendments, institution-building and common policy coordination.[39]

Reconciling from above, the founding histories of the ESRF and the European XFEL can thus be characterised as embedded into broader political contexts through their characteristics as costly and complex scientific collaborations based on intergovernmental agreement. But the interfaces between these political aspects and the manifold scientific contexts that shaped and impacted the early history of the ESRF and the European XFEL are equally important. For instance, the historical development of research with synchrotron radiation cannot be traced without dwelling into the history of particle physics because, originally, synchrotron radiation was an unwanted by-product of accelerator-based particle physics experiments. This side-note is important because the history of research with synchrotron radiation has been shaped by an uneasy relationship with particle physics research. While early research with synchrotron radiation needed to share accelerators and experimental time with particle physicists, this only changed when dedicated synchrotron radiation sources were established around the late 1980s and 1990s. Moreover, the growth of synchrotron radiation sources certainly benefited from a gradual and relative decline of large particle physics projects, most notably in the United States but also in Europe (see Chap. 3). In other words, reflections on the history of research with synchrotron radiation also need to consider paralleling historical developments in related disciplines and scientific fields. This highlights two aspects: first, that the historical developments of research with synchrotron radiation are embedded into both competitive and collaborative structures and networks among scientists, governments and national funding agencies.[40] And second, that these developments also point to longer historical trajectories of technological, scientific, political and cultural change in Europe (and the United States) since the 1950s and 1960s.

Moreover, the histories of the ESRF and the European XFEL particularly link to four different political settings that need to be briefly introduced at this point because they provide crucial backgrounds and points of reference. These contexts include, first, developments in science and technology in the United States after the end of the Second World War and the tension-laden relationship with the Soviet Union during the Cold War; second, the historical development of the bilateral relations between France and Germany from the 1960s to the mid-1980s; third, the emergence of Russia as a new player in European science and politics after the end of the Cold War; and fourth, the rocky relationship with the United Kingdom in both European politics and Big Science collaborations. Additional spotlight needs to be set on the national context of Germany, which derives from the decisive and crucial contributions of the country to the establishment of the ESRF and the European XFEL, as well as its overall powerful role in both European politics and science in recent decades.

Historians John Krige and Luca Guzzetti argued that the historical developments in the United States after the end of the Second World War constituted "a crucial point of reference to understand European big science."[41] The post-war period not only gave rise to the United States as a global military and economic power but also made it the spearhead of science and technology efforts.[42] The Manhattan Project in the 1940s represented a unique political and scientific effort on the development of nuclear weapons in the United States; born out of the fear that Germany might be able to surpass the United States by building an atomic bomb. This project paved the way for a specific relationship between science, military and the state, and demonstrated the power of science and its ability to contribute to national interests.[43] It is also widely considered to stand at the very origin of post-war Big Science.[44] This period has also been fundamental in bringing governmental patronage for basic science and Big Science, most notably related to nuclear physics, mainly because "public funding still tended to be framed in terms of arguments relating to basic research conceived as a cultural good in a free society."[45]

These developments are important to consider because they paved the way towards increasing political commitment to ever-larger high-energy/particle physics projects, and because the support of basic science in

Europe became a major concern in US-American foreign policy strategy at that time.[46] On the one hand, massive investments into ever-larger and bolder accelerator-based experiments in the United States had made this country the spearhead of high-energy/particle physics research by the 1960s, and the point of reference for European countries to catch up and compete with.[47] Five US-American national laboratories were created in the late 1940s (Argonne National Laboratory, Brookhaven National Laboratory, Lawrence Berkeley National Laboratory, Los Alamos National Laboratory and Oak Ridge National Laboratory) conducting large-scale research in the field of atomic energy and related areas such as the development of nuclear warfare.[48] By the end of the 1950s, the budget of the US-American Atomic Energy Commission (AEC) was dominated by accelerator projects in high-energy physics. At the time of the founding of Fermi National Accelerator Laboratory in the 1960s, the high-energy physics community still enjoyed a high level of financial and ideological support from the national government.[49] Investments in high-energy and nuclear physics also appeared as politically strategic investments, particularly so because the United States and the Soviet Union did not only struggle with opposing ideological systems as well as contrary foreign policy strategies and security interests. But both countries were also seeking to demonstrate superiority through the advancement of science and technology.[50] In other words, the development of the atomic bomb, the rhetorical association of nuclear power and nuclear physics with national power and security, as well as the wartime achievements of science and technology have been crucial events that made physics the crown of post-war research in the United States.[51]

On the other hand, the United States tried to intensify and strengthen its overall influence on the Western European continent. In economic terms, they feared that the absence of strong European trading partners would lead to a crisis of industrial overproduction. In political terms, they feared that Western European countries geographically close to the Soviet Union would fall to communism. The ERP (European Recovery Program), also known as the Marshall Plan, set up in 1949, has been a multilateral reconstruction programme for Western Europe. It provided a framework for European countries that financially supported reconstruction efforts in several domains.[52] This mechanism probably represents one of the most prominent examples of how Western European governments were tied to the US-American sphere of economic influence since the early post-war years.[53] But growing US-American influence also mattered, for instance, for the creation of CERN in 1954, which can be characterised as Europe's first experience in Big Science after the end of the Second World War. The setting-up of CERN as well as the activities of the US-American Rockefeller and Ford Foundation in Europe also represent cases where the United States kept an eye on the ongoing activities in post-war Europe by the means of science and technology.[54]

The 1960s and 1970s cannot only be regarded as a period of intensifying political rhetoric promoting European catch-up and competitiveness vis-à-vis the leadership of the United States in science and technology. But these decades also represent a period of deepening French-German relations. Although collaboration between the erstwhile enemies France and Germany has been fragile in the beginning of the post-war period, the agreement on the Schuman Plan and the establishment of the European Coal and Steel Community in the 1950s created a political climate in favour of collaboration. The signing of the Elysée Treaty in 1963 certainly represents a major symbolic effort of reconciliation. But previous research also illustrated that it paved the way for France and Germany to jointly establish the ILL in 1966 and to take the lead in establishing the ESRF in 1988. Over the second half of the twentieth century, the French-German tandem became a political driving force on the European continent, setting and shaping European politics as well as establishing major Big Science collaborations.[55]

British efforts to become involved in common European politics increased in the 1960s. When the domestic economic situation worsened, non-membership in the EEC apparently began to threaten British national interests and the country, as argued by historian Alexander May, "began to accept the need for membership."[56] However, British EEC membership applications were vetoed by the French president Charles de Gaulle in 1963 and 1967.[57] This situation only changed after the resignation of De Gaulle in 1969 and the nomination of Georges Pompidou as new French president. This improvement in French-British relations was paralleled by the election of the pro-European British prime minister Edward Heath in 1970. Eventually, the country joined the EEC in 1973.[58] British relations to the core countries of the EEC/EU, such as France and Germany, remained difficult, being overshadowed by a general sceptical attitude of the United Kingdom towards EU political integration.[59] With regard to the historical development of Big Science collaborations in Europe, such as CERN, ESO, ILL, EMBL, ESRF or European XFEL, negotiations with the United Kingdom often proved difficult and controversial (see Chaps. 4 and 5).[60] The United Kingdom often withdrew from projects in the midst of negotiation processes, either because of budget constraints and/or because national interests had turned to other priorities.

To the extent that the end of the Cold War as a historical turning point changed balances of power on the European continent and in the international system, it also translated into new forms of political alliances and multilateral settings. This period saw both the dissolution of the Soviet Union, Yugoslavia and Czechoslovakia, and the emergence of fourteen independent countries, as well as the re-unification of Germany as "a unique case of fusion in a decade of fission," as historian Tony Judt points out.[61] The integration of Russia as an important actor into the political and diplomatic agendas of both the individual European member countries and the EU certainly was a crucial response to this new situation in the post-Cold War. Full and formal membership of Russia in Western alliances such as the EU or NATO (North Atlantic Treaty Organization) was never seriously put on the agenda. But diplomatic relations with Russia gained new weight when the external borders of the EU were pushed closer to Russia with the enlargement round in 2004 because to "imagine a stable European political order without the inclusion of Russia in some sense would be nonsensical. Its size alone dictates a degree of inclusion."[62]

This development not only shaped post-Cold War European politics but also had an impact on Big Science in Europe. For instance, after the end of the Cold War, former member countries of the Soviet Union joined several Big Science projects in Europe, such as CERN, ILL or ESRF.[63] Russia acquired observer status at CERN in 1991, applied for associate member status in 2010 and currently negotiates a new kind of membership that "will have a much higher status and will contribute to cooperation more than associated membership."[64] The country also became a full member of the ESRF in 2014 and participates with exceptionally large financial shares in the FAIR (International Accelerator Facility for Beams of Ions and Antiprotons) and European XFEL projects (see Chap. 5).

The book is structured as follows. Chapter 2 introduces *the other Europe* as a conceptual perspective that promotes a fresh look on the development of Big Science in Europe since the second half of the twentieth century. It argues that these projects can be regarded as crucial aspects of political and scientific activities in the recent history of Europe; aspects that are, however, different from those so far foregrounded by historians, sociologists or political scientists studying European history and integration. Chapter 3 introduces and contextualises the history and science of research with synchrotron radiation from its discovery in the 1940s to current developments in the early twenty-first century. It traces the gradual development of research with synchrotron radiation, from part time use of particle physics experiments and accelerators to dedicated synchrotron radiation sources, and importantly considers competing and/or complementary developments in particle physics research and research with neutrons and ions. Chapter 4 chronicles the founding history of the ESRF, the first collaborative synchrotron radiation facility in Europe. The ESRF originated under the auspices of the European Science Foundation, but quickly escalated into a matter of high politics and intergovernmental negotiations, mainly between France, Germany and the United Kingdom. This chapter particularly highlights how the founding history of the ESRF project closely links to the strong role of the French-German tandem, in both European integration and collaborative Big Science in Europe in the 1980s. Chapter 5 investigates the founding history of the European XFEL from the 1990s to the late 2000s. It originated as a side branch of the TESLA collaboration at DESY, which had initially proposed a linear collider in particle physics. This chapter highlights important milestones during the founding phase of the European XFEL, among others, the strong role of Russia as the second biggest shareholder in this project. It further highlights the crucial role of Russia in European science and politics in the post-Cold War. Chapter 6 summarises the main findings of this book and reflects on how and to what extent the politics played out during the founding histories of the ESRF and the European XFEL stand as proxies for broader political and diplomatic concerns in Europe in the late twentieth century and the early twenty-first century. This chapter also provides an outlook to contemporary politics of Big Science in Europe updating and complementing the main historical considerations of this book.

Notes

1. See, for example, European Commission, *Developing World-Class Research Infrastructures for the European Research Area (ERA): Report of the ERA Expert Group* (Luxembourg: Office for Official Publications of the European Communities, 2008), 15; O. Hallonsten, "Research Infrastructures in Europe: The Hype and the Field." *European Review* 28, no. 4 (2020); K. C. Cramer et al. "Big Science and Research Infrastructures in Europe: History and Current Trends." In *Big Science and Research*

Infrastructures in Europe, eds. K. C. Cramer and O. Hallonsten (Cheltenham: Edward Elgar, 2020).
2. Germany here always refers to West Germany. The situation in East Germany is not discussed.
3. See, for example, H. Pollock, "The Discovery of Synchrotron Radiation." *American Journal of Physics* 51, no. 3 (1983); H. Winick and S. Doniach, "An Overview of Synchrotron Radiation Research." In *Synchrotron Radiation Research*, eds. H. Winick and S. Doniach (Boston: Springer, 1980), 4.
4. For the ESRF, see, for example, O. Hallonsten, *Small Science on Big Machines: Politics and Practices of Synchrotron Radiation Laboratories* (Lund: Research Policy Institute, 2009); O. Hallonsten, "Continuity and Change in the Politics of European Scientific Collaboration." *Journal of Contemporary European Research* 8, no. 3 (2012); V. Simoulin, *Sociologie d'un Grand Équipement Scientifique: Le Premier Synchrotron de Troisième Génération* (Lyon: ENS Éditions, 2012); H. Schmied, "The European Synchrotron Radiation Story." *Synchrotron Radiation News* 3, no. 1 (1990); H. Schmied, "The European Synchrotron Radiation Story – Phase II." *Synchrotron Radiation News* 3, no. 6 (1990). The European XFEL is mentioned in, for example, the following articles and books, but lacks comprehensive analysis: E. Lohrmann and P. Söding, *Von schnellen Teilchen und hellem Licht: 50 Jahre Deutsches Elektronen-Synchrotron DESY* (Weinheim: Wiley, 2009); T. Heinze, O. Hallonsten, and S. Heinecke, "Turning the Ship: The Transformation of DESY, 1993–2009." *Physics in Perspective* 19, no. 4 (2017); O. Hallonsten, "The Politics of European Collaboration in Big Science." In *The Global Politics of Science and Technology – Vol. 2*, eds. M. Mayer, M. Carpes and R. Knoblich (Berlin, Heidelberg: Springer, 2014).
5. See, for example, I. Ulnicane, "Ever-changing Big Science and Research Infrastructures: Evolving EU Policy." In *Big Science and Research Infrastructures in Europe*, eds. K. C. Cramer and O. Hallonsten (Cheltenham: Edward Elgar, forthcoming 2020).
6. See, for example, O. Hallonsten, H. Eriksson and A. Collsiöö, "The Role of Research Infrastructures in Innovation Systems: the Case of Swedish Participation in the Halden Reactor Project (HRP)." In *Big Science and Research Infrastructures in Europe*, eds. K. C. Cramer and O. Hallonsten (Cheltenham: Edward Elgar, forthcoming 2020); O. Hallonsten and O. Christensson, "Collaborative Technological Innovation in an Academic, User-Oriented Big Science Facility," *Industry and Higher Education* 31, no. 6 (2017): 399–408.
7. See, for example, T. Franssen, "Research Infrastructure Funding as a Tool for Science Governance in the Humanities: A Country Case Study of the

Netherlands." In *Big Science and Research Infrastructures in Europe*, eds. K. C. Cramer and O. Hallonsten (Cheltenham: Edward Elgar, forthcoming 2020); Duşa et al., eds., *Facing the Future. European Research Infrastructures for the Humanities and Social Sciences* (Berlin: Scivero Verlag, 2014).

8. See, for example, Ulnicane, "Evolving EU Policy".
9. See, for example, the contributions to K. C. Cramer and O. Hallonsten, *Big Science and Research Infrastructures in Europe* (Cheltenham: Edward Elgar, forthcoming 2020) and to A. Duşa et al., eds., *Facing the Future.*
10. See, for example, O. Hallonsten, *Big Science Transformed: Science, Politics and Organization in Europe and the United States* (Cham: Palgrave Macmillan, 2016).
11. See, for example, Hallonsten, *Big Science Transformed*; K. C. Cramer, "The Role of European Big Science in the (Geo)Political Challenges of the Twentieth and Twenty-First Centuries." In *Big Science and Research Infrastructures in Europe*, eds. K. C. Cramer and O. Hallonsten (Cheltenham: Edward Elgar, forthcoming 2020).
12. See, for example, S. Shapin, *The Scientific Life: A Moral History of a Late Modern Vocation* (Chicago: University of Chicago Press, 2008); N. Stehr, *Knowledge Societies* (London: SAGE, 1994); M. Castells and G. Cardoso, eds., *The Network Society: From Knowledge to Policy* (Washington DC: Johns Hopkins Center for Transatlantic Relations, 2005); C. Venter and D. Cohen, "The Century of Biology." *New Perspectives Quarterly* 21, no. 4 (2004); C. Kehrt, *Mit Molekülen spielen: Wissenschaftskulturen der Nanotechnologie zwischen Politik und Medien* (Bielefeld: transcript Verlag, 2016).
13. Hallonsten, *Big Science Transformed.*
14. See, for example, R. Crease and C. Westfall, "The New Big Science." *Physics Today* 69, no. 5 (2016); J. Rekers and K. Sandell, eds., *New Big Science in Focus* (Lund: Lund University Press, 2016).
15. See, for example, J. Krige, "The Politics of European Scientific Collaboration." In *Companion to Science in the Twentieth Century,* eds. J. Krige and D. Pestre (London: Routledge, 2003); J. Krige, *American Hegemony and the Postwar Reconstruction of Science in Europe* (Cambridge, MA: MIT Press, 2006); J. Krige, ed., *Choosing Big Technologies* (Chur: Harwood Academic Publishers, 1993); C. Westfall and J. Krige, "The Path of Post-War Physics." In *The Particle Century,* ed. G. Fraser (Bristol, Philadelphia: Institute of Physics Publication, 1998); P. Westwick, *The National Labs: Science in an American System, 1947–1974* (Cambridge, MA: Harvard University Press, 2003); M. Szöllösi-Janze and H. Trischler, *Großforschung in Deutschland* (Frankfurt, New York: Campus Verlag, 1990); G. Ritter, *Großforschung und Staat in Deutschland: Ein historischer*

Überblick (München: Beck, 1992); R. Seidel, “A Home for Big Science: The Atomic Energy Commission’s Laboratory System.” *Historical Studies in the Physical and Biological Sciences* 16, no. 1 (1986); P. Galison and B. Hevly, eds., *Big Science: The Growth of Large-Scale Research* (Stanford: Stanford University Press, 1992).

16. In the following, the term *common* refers to different kinds of political integration in the EEC/EU (most importantly, differentiated and uniform integration) by which at least some of the national competences where transferred and handed over to the supranational bodies and institutions of the EEC/EU.
17. See, for example, Cramer and Hallonsten, *Big Science and Research Infrastructures in Europe.*
18. See, for example, Hallonsten, *Small Science*; T. Kaiserfeld and T. O’Dell, eds., *Legitimizing ESS: Big Science as a Collaboration Across Boundaries* (Lund: Nordic Academic Press, 2013); C. Westfall, “Institutional Persistence and the Material Transformation of the US National Labs: The Curious Story of the Advent of the Advanced Photon Source.” *Science and Public Policy* 39, no. 4 (2012); P. Doing, *Velvet Revolution at the Synchrotron: Biology, Physics, and Change in Science* (Cambridge, MA: MIT Press, 2009); R. Crease, “The National Synchrotron Light Source: Part I: Bright Idea.” *Physics in Perspective* 9 (2007); R. Crease, “The National Synchrotron Light Source: Part II: The Bakeout.” *Physics in Perspective* 10 (2008).
19. See: Hallonsten, *Big Science Transformed*, 10; M. Riordan, L. Hoddeson and A. Kolb, *Tunnel Visions: The Rise and Fall of the Superconducting Super Collider* (Chicago: University of Chicago Press, 2015), ix.
20. There are also other terms such as *megascience* (see, for example, L. Hoddeson, A. Kolb, and C. Westfall, *Fermilab: Physics, the Frontier, and Megascience* (Chicago: University of Chicago Press, 2008); M. Jacob and O. Hallonsten, “The Persistence of Big Science and Megascience in Research and Innovation Policy.” *Science and Public Policy* 39, no. 4 (2012); D. Eggleton, *Examining the Relationship Between Leadership and Megascience Projects* (Doctoral Thesis: University of Sussex, 2017)) or *supersizing science* (see, for example, N. Vermeulen, *Supersizing Science: On Building Large-Scale Research Projects in Biology* (Gardners Books, 2010)).
21. See, for example, B. Hevly, “Reflections on Big Science and Big History.” In *Big Science: The Growth of Large-Scale Research*, eds. P. Galison and B. Hevly (Stanford: Stanford University Press, 1992); J. Capshew and K. Rader “Big Science: Price to the Present.” *Osiris* 7, Science after ’40 (1992); K. C. Cramer and O. Hallonsten, “Big Science and Research Infrastructures in Europe: Conclusions and Outlook.” In *Big Science and*

Research Infrastructures in Europe, eds. K. C. Cramer and O. Hallonsten (Cheltenham: Edward Elgar, forthcoming 2020).

22. Capshew and Rader, "Big Science", 22.
23. P. Papon, "Intergovernmental Cooperation in the Making of European Research." In *European Science and Technology Policy: Towards Integration or Fragmentation?* eds. H. Delanghe, U. Muldur and L. Soete (Cheltenham: Edward Elgar, 2009), 24.
24. Capshew and Rader, "Big Science".
25. See, for example, D. d. S. Price, *Little Science, Big Science* (New York: Columbia University Press, 1963); A. Weinberg, "Impact of Large-Scale Science on the United States." *Science* 134, no. 3473 (1961); A. Weinberg, *Reflections on Big Science* (Cambridge, MA: MIT Press, 1967).
26. See, for example, P. Zilsel, "The Mass Production of Knowledge." *Bulletin of the Atomic Scientists* 20, no. 4 (1964); J. Ravetz, *Scientific Knowledge and its Social Problems* (New York: Oxford University Press, 1971); P. Galison, "The Many Faces of Big Science." In *Big Science: The Growth of Large-Scale Research*, eds. P. Galison and B. Hevly (Stanford: Stanford University Press, 1992); Capshew and Rader, "Big Science".
27. High-energy physics and particle physics are used as synonyms throughout this book.
28. J. Krige, "Preface." In *Choosing Big Technologies*, ed. J. Krige (Chur: Harwood Academic Publishers, 1993), viii.
29. See, for example, Hallonsten, *Big Science Transformed*, 66; P. Galison, *Image and Logic: A Material Culture of Microphysics* (Chicago: University of Chicago Press, 1997), 671.
30. See, for example, Hallonsten, *Big Science Transformed*, 5–8.
31. Capshew and Rader, "Big Science", 12.
32. See, for instance, Krige, "The Politics of European Scientific Collaboration"; Krige, "Preface"; Hallonsten, "The Politics of European Collaboration"; Hallonsten, *Big Science Transformed*; H. Trischler and H. Weinberger, "Engineering Europe: Big Technologies and Military Systems in the Making of 20th Century Europe." *History and Technology* 21, no. 1 (2005).
33. F. Praderie, "Big Science: Why? Where? and How?" *Memorie della Società Astronomia Italiana* 67, no. 4 (1996), 898; K. C. Cramer, "Lightening Europe: Establishing the European Synchrotron Radiation Facility (ESRF)." *History and Technology* 33, no. 4 (2017), 397.
34. Exemplary Big Science projects in this regard include the role of the United States in the creation of CERN; cooperation between the United States and the Soviet Union in establishing ITER and initial ideas on the VBA (Very Big Accelerator) project.
35. See, for example, Papon, "Intergovernmental Cooperation"; P. Papon, "L'Espace Européen de la Recherche (1960–1985): Entre Science et

Politique." In *La Construction d'un Espace Scientifique Commun? La France, la RFA et l'Europe après le "Choc du Spoutnik,"* eds. C. Defrance and U. Pfeil (Bruxelles, New York: P.I.E. Peter Lang, 2012); P. Papon, "European Scientific Cooperation and Research Infrastructures: Past Tendencies and Future Prospects." *Minerva* 42, no. 1 (2004); Krige, "The Politics of European Scientific Collaboration"; Hallonsten, "The Politics of European Collaboration".

36. See, for example, Hallonsten, "The Hype and the Field".
37. See, for example, Cramer et al. "Big Science and Research Infrastructures in Europe".
38. EEC (European Economic Community) refers to the community of Europe pre-1992 (when the Maastricht Treaty was signed), and EU (European Union) refers to the same collaborative thereafter. EEC/EU and Europe are not used synonymously, but they sometimes appear side by side to emphasise that the formation and the historical development of Europe should not be conflated with the EEC/EU.
39. This approach is inspired by work of the Tensions of Europe network. See, for example, F. Schipper and J. Schot, Schipper, "Infrastructural Europeanism, or the Project of Building Europe on Infrastructures: An Introduction." *History and Technology* 27, no. 3 (2011); T. Misa and J. Schot, J. "Inventing Europe: Technology and the Hidden Integration of Europe." *History and Technology* 21, no. 1 (2005).
40. See, for example, K. Nickelsen and F. Krämer, "Introduction: Cooperation and Competition in the Sciences," *NTM Zeitschrift für Geschichte der Wissenschaften, Technik und Medizin* 24 (2016).
41. See J. Krige and L. Guzzetti, eds., *History of European Scientific and Technological Cooperation* (Luxembourg: European Communities, 1997), 439.
42. D. Pestre, "Science, Political Power and the State." In *Companion to Science in the Twentieth Century*, eds. J. Krige and D. Pestre (London: Routledge, 2003), 69.
43. See, for example, J. Hughes, *The Manhattan Project. Big Science and the Atom Bomb* (New York: Columbia University Press, 2003).
44. See, for example, C. Westfall, "Rethinking Big Science: Modest, Mezzo, Grand Science and the Development of the Bevalac, 1971–1993." *ISIS: Journal of the History of Science in Society* 94, no. 1 (2003), 33.
45. See A. Elzinga, "Features of the Current Science Policy Regime: Viewed in Historical Perspective." *Science and Public Policy* 39, no. 4 (2012), 418.
46. Krige, *American Hegemony,* 10–11.
47. See, for example, M. Lengwiler, "Kontinuitäten und Umbrüche in der Deutschen Wissenschaftspolitik des 20. Jahrhunderts." In *Handbuch Wissenschaftspolitik*, eds. S. Hornbostel, A. Knie and D. Simon (Wiesbaden:

Verlag für Sozialwissenschaften, 2010), 97, 99; Ritter, *Großforschung und Staat.*

48. See, for example, R. W. Seidel, "A Home for Big Science"; R. W. Seidel, "The National Laboratories of the Atomic Energy Commission in the early Cold War." *Historical Studies in the Physical and Biological Sciences* 32, no. 1 (2001).
49. See, for example, D. Kevles, "Cold War and Hot Physics: Science, Security, and the American State, 1945–56." *Historical Studies in the Physical and Biological Sciences* 20, no. 2 (1990); C. Westfall and L. Hoddeson, "Thinking Small in Big Science: The Founding of Fermilab, 1960–1972." *Technology and Culture* 37, no. 3 (1996).
50. See, for example, D. K. Price, *Government and Science: Their Dynamic Relation in American Democracy* (New York: NYU Press, 1954), 1.
51. See, for example, P. Forman, "Behind Quantum Electronics: National Security as Basis for Physical Research in the United States, 1940–1960." *Historical Studies in the Physical and Biological Sciences* 18, no. 1 (1987), 201.
52. See K. Middlemas, *Orchestrating Europe: The Informal Politics of the European Union 1973–1995* (London: Fontana Press, 1995), 8.
53. Middlemas, *Orchestrating Europe*, 9.
54. See, for example, K. Patel, "Rockefeller Foundation, Kalter Krieg und Amerikanisierung." In *American Foundations and the Coproduction of World Order in the Twentieth Century*, eds. H. Rausch and J. Krige (Göttingen: Vandenhoeck & Ruprecht, 2012); J. Krige, "The Ford Foundation, European Physics and the Cold War." *Historical Studies in the Physical and Biological Sciences* 29, no. 2 (1999).
55. See, for example, U. Krotz and J. Schild, *Shaping Europe: France, Germany, and Embedded Bilateralism from the Elysée Treaty to Twenty-First Century Politics* (Oxford: Oxford University Press, 2013); P. Papon, *L'Europe de la Science et de la Technologie* (Grenoble: Press Universitaires de Grenoble (PUG), 2001; M. Koopmann and J. Schild, "Eine neue Ära? Deutsch-Französische Beziehungen nach dem Ende des Kalten Krieges." In *Neue Wege in ein Neues Europa: Die Deutsch-Französischen Beziehungen nach dem Ende des Kalten Krieges*, eds. M. Koopmann, J. Schild and H. Stark (Baden-Baden: Nomos, 2013), 199; Hallonsten, *Big Science Transformed*, 89ff; C. Defrance, "France-Allemagne: Une Coopération Scientifique 'Privilégiée' en Europe, de l'Immédiat Après-Guerre au Milieu des Années 1980?" In *La Guerre Froide et l'Internationalisation des Sciences: Acteurs, Réseaux et Institutions*, eds. C. Defrance and A. Kwaschik (Paris: CNRS, 2016).
56. A. May, *Britain and Europe since 1945* (Hoboken: Taylor and Francis, 2014), 91.

57. T. Judt, *Postwar: A History of Europe since 1945* (New York et al.: Penguin Press, 2005), 292.
58. Judt, *Postwar*, 307–308, 526.
59. Judt, *Postwar*, 93.
60. See, for example, B. Jacrot, *Des Neutrons pour la Science: Histoire de l'Institut Laue-Langevin, une Coopération Internationale Particulièrement Réussie* (Les Ulis: EDP Sciences, 2006); Hallonsten, "Continuity and Change"; Cramer, "The Role of European Big Science".
61. Judt, *Postwar*, 638.
62. G. Timmins and J. Gower, "Introduction: Russia and Europe: What Kind of Partnership?" *In Russia and Europe in the Twenty-First Century: An Uneasy Partnership*, eds. J. Gower and G. Timmins (London: Anthem Press, 2009), xxii; L. Kühnhardt, *European Union - The Second Founding: The Changing Rationale of European Integration* (Baden-Baden: Nomos, 2008), 191–195.
63. For instance, after the end of Cold War, CERN was joint by Poland in 1991, the Czech Republic and Slovak Republic in 1992 and Bulgaria in 1999. The ILL was joined by the Czech Republic in 1999, by Hungary in 2005, by Poland in 2006 and by Slovakia in 2009.
64. Joint Institute for Nuclear Research, "Russia and CERN are Working Out a New Format of Cooperation," News Release (14 March 2018).

Bibliography

Bush, V. *Science: The Endless Frontier*. Washington: Government Printing Office, 1945.

Capshew, J. and Rader, K. "Big Science: Price to the Present." *Osiris* 7, Science after '40 (1992): 2–25.

Castells, M. and Cardoso, G., eds. *The Network Society: From Knowledge to Policy*. Washington DC: Johns Hopkins Center for Transatlantic Relations, 2005.

Cramer, K. C. "The Role of European Big Science in the (Geo)Political Challenges of the Twentieth and Twenty-First Centuries." In *Big Science and Research Infrastructures in Europe*, edited by K. C. Cramer and O. Hallonsten, 56–75. Cheltenham: Edward Elgar, 2020.

Cramer, K. C. "Lightening Europe: Establishing the European Synchrotron Radiation Facility (ESRF)." *History and Technology* 33, no. 4 (2017): 396–427.

Cramer, K. C. and Hallonsten, O. "Big Science and Research Infrastructures in Europe: Conclusions and Outlook." In *Big Science and Research Infrastructures in Europe*, edited by K. C. Cramer and O. Hallonsten, 251–257. Cheltenham: Edward Elgar, 2020.

Cramer, K. C., Hallonsten, O., Bolliger, I., and Griffiths, A. "Big Science and Research Infrastructures in Europe: History and Current Trends." In *Big*

Science and Research Infrastructures in Europe, edited by K. C. Cramer and O. Hallonsten, 1–26. Cheltenham: Edward Elgar, 2020.

Crease, R. "The National Synchrotron Light Source: Part I: Bright Idea." *Physics in Perspective* 9 (2007): 1–30.

Crease, R. "The National Synchrotron Light Source: Part II: The Bakeout." *Physics in Perspective* 10 (2008): 1–31.

Crease, R. and Westfall, C. "The New Big Science." *Physics Today* 69, no. 5 (2016): 30–36.

Defrance, C. "France-Allemagne: Une Coopération Scientifique 'Privilégiée' en Europe, de l'Immédiat Après-Guerre au Milieu des Années 1980?" In La Guerre Froide et l'Internationalisation des Sciences: Acteurs, Réseaux et Institutions. Edited by C. Defrance and A. Kwaschik, 169–186. Paris: CNRS, 2016.

Doing, P. *Velvet Revolution at the Synchrotron: Biology, Physics, and Change in Science*. Cambridge, MA: MIT Press, 2009.

Duşa, A., Nelle, D., Stock, G. and Wagner G. G., eds., *Facing the Future. European Research Infrastructures for the Humanities and Social Sciences.* Berlin: Scivero Verlag, 2014.

Eggleton, D. *Examining the Relationship Between Leadership and Megascience Projects.* Doctoral Thesis: University of Sussex, 2017.

Elzinga, A. "Features of the Current Science Policy Regime: Viewed in Historical Perspective." *Science and Public Policy* 39, no. 4 (2012): 416–28.

European Commission. *Developing World-class Research Infrastructures for the European Research Area (ERA): Report of the ERA Expert Group.* Luxembourg: Office for Official Publications of the European Communities, 2008.

Forman, P. "Behind Quantum Electronics: National Security as Basis for Physical Research in the United States, 1940–1960." *Historical Studies in the Physical and Biological Sciences* 18, no. 1 (1987): 149–229.

Franssen, T. "Research Infrastructure Funding as a Tool for Science Governance in the Humanities: A Country Case Study of the Netherlands." In *Big Science and Research Infrastructures in Europe*, edited by K. Cramer and O. Hallonsten, 157–176. Cheltenham: Edward Elgar, 2020.

Galison, P. "The Many Faces of Big Science." In *Big Science: The Growth of Large-Scale Research*, edited by P. Galison and B. Hevly, 1–17. Stanford: Stanford University Press, 1992.

Galison, P. and Hevly, B., eds., *Big Science: The Growth of Large-Scale Research.* Stanford: Stanford University Press, 1992.

Galison, P. *Image and Logic: A Material Culture of Microphysics.* Chicago: University of Chicago Press, 1997.

Hallonsten, O. "Continuity and Change in the Politics of European Scientific Collaboration." *Journal of Contemporary European Research* 8, no. 3 (2012): 300–319.

Hallonsten, O. "Research Infrastructures in Europe: The Hype and the Field." *European Review* 28, no. 4 (2020): 617–635.

Hallonsten, O. "The Parasites." *Historical Studies in the Natural Sciences* 45, no. 2 (2015): 217–272.

Hallonsten, O. "The Politics of European Collaboration in Big Science." In *The Global Politics of Science and Technology – Vol. 2*. Edited by M. Mayer, M. Carpes and R. Knoblich, 31–46. Berlin, Heidelberg: Springer, 2014.

Hallonsten, O. *Big Science Transformed: Science, Politics and Organization in Europe and the United States*. Cham: Palgrave Macmillan, 2016.

Hallonsten, O. *Small Science on Big Machines: Politics and Practices of Synchrotron Radiation Laboratories*. Lund: Research Policy Institute, 2009.

Hallonsten, O., Eriksson, H., and Collsiöö, A. "The Role of Research Infrastructures in Innovation Systems: The Case of Swedish Participation in the Halden Reactor Project (HRP)." In *Big Science and Research Infrastructures in Europe*, edited by K. Cramer and O. Hallonsten, 177–197. Cheltenham: Edward Elgar, 2020.

Heinze, T., Hallonsten, O., and Heinecke, S. "Turning the Ship: The Transformation of DESY, 1993–2009." *Physics in Perspective* 19, no. 4 (2017): 424–451.

Hevly, B. "Reflections on Big Science and Big History." In *Big Science: The Growth of Large-Scale Research*, edited by P. Galison and B. Hevly, 355–364. Stanford: Stanford University Press, 1992.

Hoddeson, L., Kolb, A. and Westfall, C. *Fermilab: Physics, the Frontier, and Megascience*. Chicago: University of Chicago Press, 2008.

Jacob, M. and Hallonsten, O. "The Persistence of Big Science and Megascience in Research and Innovation Policy." *Science and Public Policy* 39, no. 4 (2012): 411–415.

Joint Institute for Nuclear Research. "Russia and CERN are Working Out a New Format of Cooperation," News Release, 14 March 2018. Online available: http://www.jinr.ru/posts/russia-and-cern-are-working-out-a-new-format-of-cooperation, last accessed 20 March 2020.

Judt, T. *Postwar: A History of Europe since 1945*. New York et al.: Penguin Press, 2005.

Kaiserfeld, T. and O'Dell, T., eds. *Legitimizing ESS: Big Science as a Collaboration Across Boundaries*. Lund: Nordic Academic Press, 2013.

Kehrt, C. *Mit Molekülen spielen: Wissenschaftskulturen der Nanotechnologie zwischen Politik und Medien* (Bielefeld: transcript Verlag, 2016).

Kevles, D. "Cold War and Hot Physics: Science, Security, and the American State, 1945–56." *Historical Studies in the Physical and Biological Sciences* 20, no. 2 (1990): 239–264.

Koopmann, M. and Schild, J. "Eine neue Ära? Deutsch-Französische Beziehungen nach dem Ende des Kalten Krieges." In *Neue Wege in ein Neues Europa: Die*

Deutsch-Französischen Beziehungen nach dem Ende des Kalten Krieges, edited by M. Koopmann, J. Schild and H. Stark, 199–210. Baden-Baden: Nomos, 2013.

Krige, J., ed. *Choosing Big Technologies*. Chur: Harwood Academic Publishers, 1993a.

Krige, J. "Preface." In *Choosing Big Technologies*, edited by J. Krige, vii–xii. Chur: Harwood Academic Publishers, 1993b.

Krige, J. "The Politics of European Scientific Collaboration." In *Companion to Science in the Twentieth Century*, edited by J. Krige and D. Pestre, 897–918. London: Routledge, 2003.

Krige, J. *American Hegemony and the Postwar Reconstruction of Science in Europe*. Cambridge, MA: MIT Press, 2006.

Krige, J. and Guzzetti, L., eds. *History of European Scientific and Technological Cooperation*. Luxembourg: European Communities, 1997.

Krotz, U. and Schild, J. *Shaping Europe: France, Germany, and Embedded Bilateralism from the Elysée Treaty to Twenty-First Century Politics*. Oxford: Oxford University Press, 2013.

Kühnhardt, L. *European Union – The Second Founding: The Changing Rationale of European Integration*. Baden-Baden: Nomos, 2008.

Lengwiler, M. "Kontinuitäten und Umbrüche in der Deutschen Wissenschaftspolitik des 20. Jahrhunderts." In *Handbuch Wissenschaftspolitik*, edited by S. Hornbostel, A. Knie and D. Simon, 6–18. Wiesbaden: Verlag für Sozialwissenschaften, 2010.

Lohrmann, E. and Söding, P. *Von schnellen Teilchen und hellem Licht: 50 Jahre Deutsches Elektronen-Synchrotron DESY*. Weinheim: Wiley, 2009.

May, A. *Britain and Europe since 1945*. Hoboken: Taylor and Francis, 2014.

Middlemas, K. *Orchestrating Europe: The Informal Politics of the European Union 1973–1995*. London: Fontana Press, 1995.

Misa, T. and Schot, J. "Inventing Europe: Technology and the Hidden Integration of Europe." *History and Technology* 21, no. 1 (2005): 1–19.

Nickelsen, K. and Krämer, F. "Introduction: Cooperation and Competition in the Sciences," *NTM Zeitschrift für Geschichte der Wissenschaften, Technik und Medizin* 24 (2016): 119–123.

Papon, P. "European Scientific Cooperation and Research Infrastructures: Past Tendencies and Future Prospects." *Minerva* 42, no. 1 (2004): 61–76.

Papon, P. "Intergovernmental Cooperation in the Making of European Research." In *European Science and Technology Policy: Towards Integration or Fragmentation?* edited by H. Delanghe, U. Muldur, and L. Soete, 24–43. Cheltenham: Edward Elgar, 2009.

Papon, P. "L'Espace Européen de la Recherche (1960–1985): Entre Science et Politique." In *La Construction d'un Espace Scientifique Commun? La France, la RFA et l'Europe après le "Choc du Spoutnik"*. Edited by C. Defrance and U. Pfeil, 37–54. Bruxelles, New York: P.I.E. Peter Lang, 2012.

Papon, P. *L'Europe de la Science et de la Technologie*. Grenoble: Press Universitaires de Grenoble (PUG), 2001.

Patel, K. "Rockefeller Foundation, Kalter Krieg und Amerikanisierung." In *American Foundations and the Coproduction of World Order in the Twentieth Century*, edited by H. Rausch and J. Krige, 173–185. Göttingen: Vandenhoeck & Ruprecht, 2012.

Pestre, D. "Science, Political Power and the State." In *Companion to Science in the Twentieth Century*. Edited by J. Krige and D. Pestre, 61–75. London: Routledge, 2003.

Pollock, H. "The Discovery of Synchrotron Radiation." *American Journal of Physics* 51, no. 3 (1983): 278–280.

Praderie, F. "Big Science: Why? Where? and How?" *Memorie della Società Astronomia Italiana* 67, no. 4 (1996): 889–899.

Price, D. K. *Government and Science: Their Dynamic Relation in American Democracy*. New York: NYU Press, 1954

Price, Derek J. de Solla. *Little Science, Big Science*. New York: Columbia University Press, 1963.

Ravetz, J. *Scientific Knowledge and its Social Problems*. New York: Oxford University Press, 1971.

Rekers, J. and Sandell, K., eds. *New Big Science in Focus*. Lund: Lund University Press, 2016.

Riordan, M., Hoddeson, L. and Kolb, A. *Tunnel Visions: The Rise and Fall of the Superconducting Super Collider*. Chicago: University of Chicago Press, 2015.

Ritter, G. *Großforschung und Staat in Deutschland: Ein historischer Überblick*. München: Beck, 1992.

Schipper, F. and Schot, J. "Infrastructural Europeanism, or the Project of Building Europe on Infrastructures: An Introduction." *History and Technology* 27, no. 3 (2011): 245–264.

Schmied, H. "The European Synchrotron Radiation Story – Phase II." *Synchrotron Radiation News* 3, no. 6 (1990a): 22–26.

Schmied, H. "The European Synchrotron Radiation Story." *Synchrotron Radiation News* 3, no. 1 (1990b): 18–22.

Seidel, R. W. "A Home for Big Science: The Atomic Energy Commission's Laboratory System." *Historical Studies in the Physical and Biological Sciences* 16, no. 1 (1986): 135–175.

Seidel, R. W. "The National Laboratories of the Atomic Energy Commission in the early Cold War." *Historical Studies in the Physical and Biological Sciences* 32, no. 1 (2001): 145–162.

Shapin, S. *The Scientific Life: A Moral History of a Late Modern Vocation*. Chicago: University of Chicago Press, 2008.

Simoulin, V. *Sociologie d'un Grand Équipement Scientifique: Le Premier Synchrotron de Troisième Génération*. Lyon: ENS Éditions, 2012.

Stehr, N. *Knowledge Societies*. London: SAGE, 1994.
Szöllösi-Janze, M. and Trischler, H. *Großforschung in Deutschland*. Frankfurt, New York: Campus Verlag, 1990.
Timmins, G. and Gower, J. "Introduction: Russia and Europe: What Kind of Partnership?" *In Russia and Europe in the Twenty-First Century: An Uneasy Partnership*. Edited by J. Gower and G. Timmins, xxi-xxvi. London: Anthem Press, 2009.
Trischler, H. and Weinberger, H. "Engineering Europe: Big Technologies and Military Systems in the Making of 20th Century Europe." *History and Technology* 21, no. 1 (2005): 49–83.
Ulnicane, I. "Ever-changing Big Science and Research Infrastructures: Evolving EU Policy." In *Big Science and Research Infrastructures in Europe*, edited by K. Cramer and O. Hallonsten, 76–100. Cheltenham: Edward Elgar, 2020.
Venter, C. and Cohen, D. "The Century of Biology." *New Perspectives Quarterly* 21, no. 4 (2004): 73–77.
Vermeulen, N. *Supersizing Science: On Building Large-Scale Research Projects in Biology*. Gardners Books, 2010.
Weinberg, A. "Impact of Large-Scale Science on the United States." *Science* 134, no. 3473 (1961): 161–164.
Weinberg, A. *Reflections on Big Science*. Cambridge, Mass.: MIT Press, 1967.
Westfall, C. "Institutional Persistence and the Material Transformation of the US National Labs: The Curious Story of the Advent of the Advanced Photon Source." *Science and Public Policy* 39, no. 4 (2012): 439–449.
Westfall, C. "Rethinking Big Science: Modest, Mezzo, Grand Science and the Development of the Bevalac, 1971–1993." *ISIS: Journal of the History of Science in Society* 94, no. 1 (2003): 30.
Westfall, C. and Hoddeson, L. "Thinking Small in Big Science: The Founding of Fermilab, 1960–1972." *Technology and Culture* 37, no. 3 (1996): 457–492
Westfall, C. and Krige, J. "The Path of Post-War Physics." In *The Particle Century*. Edited by G. Fraser, 1–11. Bristol, Philadelphia: Institute of Physics Pub., 1998.
Westwick, P. *The National Labs: Science in an American System, 1947–1974*. Cambridge, MA: Harvard University Press, 2003.
Winick, H., Doniach, S. "An Overview of Synchrotron Radiation Research." In *Synchrotron Radiation Research*, edited by H. Winick and S. Doniach, 1–10. Boston, MA: Springer, 1980.
Zilsel, P. "The Mass Production of Knowledge." *Bulletin of the Atomic Scientists* 20, no. 4 (1964): 28–29.

CHAPTER 2

What Kind of Europe for European Big Science?

"The term 'European cooperation,'" as argued by historian Corine Defrance, "encompasses extremely varied situations. Because the Europe of research has been built on different scales, on a geographical and political basis that is broader and more diversified than that of the European community."[1] Big Science projects are a specific kind of "European cooperation" in the sense that these projects throughout the second half of the twentieth century and the early twenty-first century remain to be based on intergovernmental agreement and *formally* disentangled from common[2] EEC/EU politics and policymaking.[3] The EEC/EU only had a marginal role to play within Big Science activities in Europe, mainly due to a lack of a coherent and common research policy. The EEC/EU acted, first and foremost, as a funder through its Framework Programmes (Framework Programmes for Research and Technological Development, FP) that were established in 1984. The FPs started to support researcher mobility among Big Science facilities in Europe in the late 1999s and began to pledge funding to preparatory phases or upgrade programmes of these projects in the early 2000s.[4] But the EEC/EU had little to say about how these collaborations were established, implemented and organised.

The relationship between the EEC/EU and Big Science collaborations in Europe is, however, more complex than it might appear on first sight. It can be observed that to the extent that Big Science collaborations are formally disentangled from common EU politics and policymaking, they are at the same time deeply enmeshed with patterns and dynamics of

K. C. Cramer, *A Political History of Big Science*,
Palgrave Studies in the History of Science and Technology,
https://doi.org/10.1007/978-3-030-50049-8_2

European politics and diplomacy. In other words, Big Science is built on a different organisational and political scale than the EEC/EU framework. However, and this is one of the core messages of this book, patterns of diplomatic and political relations among countries in and around Europe, moments of deepening European integration, times of European political crisis and upheavals are nevertheless well resonated by the politics played out during the founding phases of Big Science collaborations. This pattern is remarkable because the specific character of intergovernmental Big Science projects—being formally independent from broader EEC/EU frameworks—seemingly makes them stand out in the recent history of Europe and the EU. There is hence much to suggest that Big Science projects constitute complementary pieces in the European integration puzzle and in the assessment of the spatial and political limits of Europe that sheds fresh light on European multilateral politics, alliance-building and integration dynamics. They can be regarded as crucial aspects of political and scientific activities in the recent history of Europe; aspects that are, however, different from those so far foregrounded by historians, sociologists or political scientists studying European history and integration. Through the conceptual perspective of *the other Europe*, this book seeks to close this gap in scholarly understanding, namely how the relationship between, on the one hand, Europe and the EEC/EU as well as, on the other hand, Big Science projects can be characterised and how this improves our knowledge of both the politics of Big Science and the history of Europe and the EEC/EU.

2.1 The Other Europe

The other Europe is a conceptual perspective that promotes a fresh look on the development of Big Science in Europe since the second half of the twentieth century to improve scholarly knowledge of the interfaces of science and politics within large and single-sited scientific collaborations. *The other Europe* draws its main inspiration from three distinct, yet related, perspectives on the recent history of Europe (see below). In doing so, it particularly emphasises three different dimensions, namely technology, spatiality and politics that are deemed to be of major significance to gain a nuanced understanding of the historical and contemporary dynamics of Big Science collaborations.

First, *the other Europe* borrows from multifaceted scholarly work within the Tensions of Europe (ToE)[5] network on the roles of infrastructures and (large[6]) technologies in making and shaping Europe, European politics and the European integration process. It challenges a traditional research canon that merely focused on treaty reforms and amendments, institution-building and policy coordination as the dominant picture of the recent history of Europe and the EEC/EU.[7] By neither collapsing European integration into a teleological process nor attributing ultimate explanatory power to either political or technological aspects, the manifold perspectives from scholars of the ToE network offer a workable starting point for the conceptual framework of this book to rely on, as well as for Big Science collaborations in Europe to be studied.[8]

Second, *the other Europe* also relates to the work of historian Kiran Klaus Patel who similarly promoted "to rethink the way we write European integration history."[9] Patel criticised that historiography and interdisciplinary scholarship on European integration too often centred on internal institutional mechanisms and common policy principles. But it missed to consider "the various routes of exchanges" and to embed the historical developments into broader contexts of "European, trans-Atlantic and global co-operation."[10] Relating to Patel, *the other Europe* tries to shed light on the science-politics interfaces of Big Science collaborations by highlighting the many connections between the different local, national, transnational and European contexts, under which the projects are negotiated and realised.

Third, *the other Europe*, more generally, also draws on broader frameworks in sociology and history that regard science and politics as "mutual resource systems"[11] to explore the interfaces between science and politics and to investigate and understand their relations and dynamics.[12] Such a perspective, as argued by historian Helmuth Trischler, also "allows the historian to study the changing interdependencies of science, politics, economy, military and the public over time."[13] With regard to the specific topic of Big Science, such a conceptual perspective is particularly represented through the works of the historians John Krige and Dominique Pestre, former science administrator Pierre Papon or sociologist Olof Hallonsten.[14]

Summarising from above, *the other Europe* relates to all three perspectives in that it characterises collaborative Big Science projects as crucial, but largely overlooked political and scientific links between several countries that contribute, as objects and mediums of investigation, to an enriched perspective on the history of Europe. More concretely, it conceptualises these projects as empirical examples of the dynamics of European politics, diplomacy and political integration that considerably enrich our understanding of how Europe was performed and projected at different times.

2.1.1 *Technology*

Technology is a core concern for the conceptual perspective of *the other Europe*. Several scholars within the ToE network characterised technology "not only as machines, products, systems, and infrastructures, but also as the skills and knowledge that make them work."[15] Emphasis was put, among other topics, on information and communication technologies, transport technologies, consumer technologies and "knowledge intensive technologies of large-scale European projects."[16] The importance of technology for Big Science projects has often been overlooked in scholarly research because Big Science, literally, puts emphasis on the science. However, historians James Capshew and Karen Rader argued that technology is "a hallmark, if not the main driving force behind Big Science."[17] Hence, this book characterises technology as the ensemble of materials, instruments and/or tools that provide the necessary resources to carry out specific experiments at Big Science facilities. Technology also includes "skills, knowledge, experience, together with suitable organizational and institutional arrangement"[18] (what innovation scholar Richard Li-Hua summarises as *software*)[19] to be constructed and operated, which practically translates into large teams of, for example, technicians, scientists and administrators at Big Science facilities, as well as into complex organisational and administrative frameworks.

Technology also plays a decisive role for the ESRF and the European XFEL by providing the main material baseline around which the history and politics of the founding phases of the two facilities circled. Both facilities are accelerator-based light sources. The accelerator complex constitutes the material basis for the light to be produced, used and exploited, that is, for the science and the experiments at these facilities to be done (see Chap. 3). It also arranges several smaller technologies such as magnets or vacuum tubes as well as knowledge and expertise through, for instance, the work of accelerator physicists or beamline scientists into an operational light source. These technologies are not only large but also costly and complex, which means that these Big Science efforts are hardly duplicated at any other point in Europe and beyond.

Science scholar Sheila Jasanoff highlighted these multifaceted connections under the umbrella term of *co-production* adding emphasis on the political and economic contexts in which technologies and scientific work are placed: "The design of technology is likewise seldom accidental; it reflects the imaginative faculties, cultural preferences and economic or political resources of their makers and users."[20] This conceptual statement is paralleled by empirical observation, namely, that political commitment, design and approval of new Big Science projects often refrain contemporary science policy rationales and economic calculation. Particularly with regard to the European XFEL and similar new Big Science projects, such as FAIR (Facility for Antiproton and Ion Research), that were constructed in the early 2010s, the technologies of these projects received considerable political interest. University departments and research institutes increasingly specialised in accelerator development; governmental representatives and funding agencies began to consider technologies as the innovative hubs within Big Science projects from which promising industrial spin-offs, patents and commercial applications were expected to emerge.[21]

Summarising the above, large and complex technologies are not only fundamental but also visible characteristics of single-sited Big Science collaborations. To the extent that recent decades have also witnessed the emergence of several types of distributed Research Infrastructures, including cloud-based ones, that can be operated from practically any point around the world, several Big Science projects (including the ESRF and the European XFEL) remain to be single-sited: They host large technologies, particularly in the natural sciences, that are unable to move. With regard to the importance of their performances for the advancement of science together with the unavoidable localisation of their technologies at specific sites, they can hence be characterised as precious but also crucial resources in the European scientific landscapes.

2.1.2 Spatiality

Relating to the above said, research activities within the ToE network extensively investigated how technologies made and shaped Europe asking "how technical communities, companies, nation states and social groups have contested, projected, performed, and reproduced 'Europe' in constructing and using a range of technologies."[22] In this context, it importantly needs to be considered that the members of this network also proposed to study "Europe as something more than a collection of partly contrasting and partly overlapping national experiences."[23] For instance, historians Frank Schipper and Johan Schot emphasise the simultaneous co-construction of different Europes regarding flexible interpretations and conceptions of scope, content and meaning "to prevent the conflation of the formation of Europe with that of the formal process of European integration."[24] In other words, today's EU is but one specific projection of Europe. This observation also applies to Big Science in Europe that is both formally disentangled from common EU policies, but closely connected to European politics. With regard to the ESRF, the European XFEL and the omnipresence of the notion *European* in the official names of many other Big Science projects,[25] it can be speculated that their understandings of what Europe and the EEC/EU are and how they frame the spatial and political limits of Europe differ. Related to this, there is hence much to suggest that through and by the historical study of Big Science projects in Europe many different conceptions and interpretations of Europe are waiting to be studied and revealed.

The above-mentioned conceptual framing of Europe contrasts and challenges, on the one hand, a purely state-centric perspective and adds, on the other hand, a specific transnational dimension. *Transnational*, according to historian Erik van der Vleuten, mainly acquires three different meanings: the investigation of transnational infrastructures or cross-border flows, the study of international organisations (governmental and nongovernmental), as well as those research topics that cross and/or run counter well-established analytical categories such as national or local contexts.[26] Similarly, historian John Krige characterised a launcher experiment by two French space scientists in the late 1950s as a "global hybrid."[27] His argument aimed to reveal how supposedly national space activities are "embedded in a network of interconnections and overflowed territorial boundaries,"[28] and how a comprehensive analysis of such and similar cases in the history of science and technology aim to "rupture the national frame and to situate scientific and technological practices in a transnational or global framework."[29]

Big Science collaborations in Europe can equally be characterised as transnational endeavours, following van der Vleuten, and a European "hybrid," following John Krige. Their transnational character mainly springs from the tensions that arise between their local anchoring as single-sited facilities and their embeddedness into broader political and scientific contexts through their characteristics as publicly funded, intergovernmental projects, as well as through the transborder relationships of their materials, in-house scientists, external researchers, administrators and governmental representatives. In other words, to the extent that the ESRF and the European XFEL are single-sited, they are geographically identifiable locations, where scientific practice and materiality are physically co-located. But these projects also need to consider questions of funding, benefit, (scientific) fair return, participation and contribution that are all politics and that open to broader contexts beyond the local one (see Sect. 2.1.3 of this chapter).

While it has been highlighted above that the local context is of particular importance for single-sited Big Science facilities, John Krige importantly adds that we should not

> blur the role of national actors in writing transnational history. Our task rather is to recapture the not-so-nebulous web of international linkages that are mobilized by national actors to serve national and foreign policy objectives.[30]

Krige further argues that "[c]ollaboration, then, and European scientific collaboration in particular, is not undertaken at the expense of self-interest; it is rather, the pursuit of one's interest by other means."[31] Following John Krige, there is much to suggest that Big Science collaborations in Europe carry a strong national presence and that national interests and strategies matter greatly to commit and support collaborative efforts in science. The intergovernmental framework of Big Science collaborations partly explains the strong role of national agendas: Intergovernmental agreements still come into being (through) informal meetings, negotiations and discussions among scientists, administrators and governmental representatives. Controversy caused by diverging national agendas and priorities often paralled compromise on the common good. Bargaining on a site, financial contributions and voting procedures that were recurrent issues during the founding phases of many collaborative Big Science facilities in Europe testify of usual power plays within these and similar multilateral settings (see also Sect. 2.1.3 of this chapter and Chaps. 4 and 5). But it also needs to be considered that such an intergovernmental framework offers the possibility for the individual collaborating countries to keep the framework, scope and contours of the project under their power and control. It also allows them to bring in national agendas and priorities and, to put it simply, to remain in the driving seat of this cost-intensive, long-term and often publicly visible and prestigious effort as far as possible and as long as possible.

However, intergovernmental Big Science projects certainly escalate beyond a simple juxtaposition of different national narratives. Exclusive emphasis on the roles and strategies of governmental representatives and national frames are problematic in the sense that the dynamics of Big Science cannot be entirely explained by these aspects. With regard to a global space history, John Krige argues that it "must retain the national as a key analytical category – not as an autonomous but as an interdependent actor, whose scientific and technological practices are inspired by national interest and framed by foreign policy."[32] In other words, an ontological dichotomy between national and transnational developments and contexts, as well as the territorial representation of the world made by discrete sovereign nation-states, neither does constitute the full spectrum of what *the other Europe* is nor does determine the role of Big Science collaborations in Europe with regard to European politics, European integration and the recent past of Europe. The transnational setting, in this regard, should hence not be understood as a rebellion against the state-centric view of world affairs.[33] A transnational perspective rather highlights a multifaceted compendium of cross-border relations that play out during the establishment, construction and operation of these Big Science projects, and that also include emphasis on the role of national governments and the national context. These single-sited, intergovernmental Big Science collaborations hence carry local, national as well as specific transnational dimensions, which connect and assemble different actors in science and politics. This leads to the formation of a historical narrative that contains specific representations of what Europe meant and how it was represented at different times.

2.1.3 Politics

Big Science collaborations in Europe need to gather political support and commitment from several countries to be funded and realised. This closely connects to considerations in the previous section, namely, that the national context and emphasis on the state/government as powerful actor remain crucial aspects in the history and politics of Big Science. To the extent that national agendas, strategies and interests shape the establishment of Big Science collaborations and that intergovernmental agreements needs to be reached on location, financial shares and the legal framework, politics come to matter.[34] Politics, in this context, is understood as the ability to shape and frame outcomes of intergovernmental and multilateral negotiations through bargaining, package-deal creation or alliance-building, and the power to define the contours and directions of future Big Science projects in Europe. The strong national presence can be a source of both political disarray and rapid decision-making for these projects. On the one hand, as a variable consortium of countries joins and commits to these projects, much room exists for choosing close partners, and for informally concluding bilateral agreements and decisions. As will be discussed in Chaps. 4 and 5, bilateral alliances between France and Germany as well as Germany and Russia were decisive for the ESRF and European XFEL project to advance and to be established. On the other hand, the way science and research are organised in different countries varies significantly, and so do national research policy agendas and strategies, which can all be a source of conflict and delay for the project; particularly so with regard to the formation of a common EU research policy (see also Sect. 2.2 of this chapter).

But the question is: what are the reasons and motivations for national governments to initiate or join such collaborative efforts? Previous scholarly research revealed mainly three different aspects: First, the sharing of costs saves national resources and opens a large room for manoeuvre when costs increase unpredictably or when damage calls for urgent and costly action. This was remarked by historian Corine Defrance, among others, with regard to the situation of energy and finance crisis in Europe in the 1960s and 1970s.[35] Second, scientific collaboration was also regarded as a way to acquire a stronger joint European voice, first and foremost with regard to the dominating role of the United States after the end of the Second World War and during the Cold War.[36] Third, collaboration, and Big Science collaboration in particular, also served as an identificatory moment for European countries in times of political crisis and disarray.[37] Yet, it remains questionable whether these and similar aspects were important enough for countries to initiate or to engage in collaborative efforts in Big Science because political commitment to these projects also means to tie several countries and part of their national science budgets together over a very long period. John Krige, for instance, highlights that to catch up with US-American leadership in science and technology in the first decades after the end of the Second World War appeared quite impossible due to the enormous sums spent by the United States on science and technology. "Competition was," as further argued by John Krige "out of the question in the short to medium term. The aim both in Western Europe and in Japan, was rather autonomy in the longer term."[38] Moreover, John Krige also argues that solo efforts in Big Science were always a plausible option for the larger European countries such as France, (West) Germany or the United Kingdom.[39]

Summarising from above, it seems as if political calculation prevailed over any other kind of consideration namely that Big Science in Europe was and remains not much an issue of shared intrinsic motivation and joining efforts for the common good. Drawing on insights from previous research, there is much to suggest that political calculation of duties and benefits, of tit-for-tat-strategies on national priorities and of broader package deals made countries collaborate and impacted the founding phases of major European Big Science facilities.[40] It needs, however, also to be considered that despite this apparently dominating role of national interests, governments nevertheless decided to join forces on a long-term basis to establish, construct and operate cost-intensive Big Science facilities. Collaboration and self-interests, so it seems, were successfully played, which may provide a part of the answer to why collaborative Big Science projects were established in Europe, while being largely absent from other parts of the world.

Related to the above, the dynamics of intergovernmental (dis)agreement and the pursuit of national interests also have severe consequences on how Big Science projects are negotiated, agreed and perceived, as well as on the role(s) that they are expected to play in the European (political) arena. It can further be argued that the establishment of collaborative Big Science projects provided important identificatory moments for the making and shaping of Europe to the extent that membership and non-membership also represent particularly (geo)political projections of Europe.[41] This partly contradicts what Luca Guzzetti, Pierre Papon and others have discussed under the heading of a "variable geometry" and/or the "Europe à la carte" model for intergovernmental Big Science collaborations in Europe.[42] The term was borrowed from the discourse on differentiated European integration, including, for instance, integration by a variable number of countries, as it is the case for the Schengen Agreement, the Euro zone or the EMU (Economic and Monetary Union).[43] According to Guzzetti, Papon and others, a variable number of countries (members of the EEC/EU as well as non-members) voluntarily commit to such initiatives to jointly fund and support the realisation of a research project. While this definition describes the patterns that stand behind intergovernmental Big Science projects in Europe quite well, the *voluntary* nature of joining intergovernmental Big Science projects, as described above, clearly is an ideal type that should not always be taken for granted.

On the one hand, the many Big Science facilities that were established in Europe over the course of the last decades has created a net of obligations, duties and benefits that can also pressure countries to always participate in the next-generation project. In this regard, participation in future efforts becomes a self-fulfilling prophecy, if perceived as a means to balance duties and benefits from past engagements.[44] On the other hand, pressure has also been put on the smaller, scientifically as well as politically less powerful countries because, for them, collaborative efforts often constitute the only possibility to access cutting-edge research. In order to keep path with development within larger (European) countries, participation in collaborative efforts for them seems as a must.[45]

As already mentioned, politics are crucial for the founding phases of collaborative Big Science projects and manifest, not exclusively, but mainly through two aspects: site selection and financial shares. The site selection process seldom remains a (simple) question of geography. But it also involves national interests and political strategies, as well as a diverging set of backgrounds, knowledge and intentions. From a scientist's perspective, it needs, for instance, to consider the limitations of the physical world, such as the availability of cooling water or the condition of the soil, as well as seismic activities for tunnelling and installing heavy instruments. From the point of view of governmental representatives, harsh negotiations over location often lead to the agreement on a site premium to be paid by the hosting country that should compensate the many (economic) benefits of having the facility built on its own national territory.

Probably each single Big Science collaboration has its own story of site selection to tell; and so do the ESRF and the European XFEL projects (see Chaps. 4 and 5).[46] Several examples in the history of Big Science testify of the many difficulties of site selection. The amendment of the original CERN (European Organization for Nuclear Research) facility in the 1970s (CERN II) and the efforts to reach agreement on an appropriate site caused intergovernmental disagreement. After West Germany and the United Kingdom temporally withdrew from the project, this situation could be solved only by locating CERN II on the same site as CERN in the 1950s, that is (politically) neutral Switzerland.[47] Similarly, the United Kingdom withdrew from ESO (European Southern Observatory) in 1960 because it preferred to support another astronomical project to be located in Australia, as part of the British Commonwealth.[48]

Another aspect is negotiations of financial shares. As it is a common principle for publicly funded Big Science facilities, their use by the scientific user communities is based on scientific criteria. It does not come as a surprise that countries are reluctant to financially support a collaborative facility that would not be built on their own territory. One consequence of this is that

> countries have little or no reason to enter the collaboration a level higher than absolutely necessary or possible (…), because their scientific communities will have access to the facility anyway, to the extent that they can compete with scientific quality.[49]

Governments and funding agencies often looked for ways to maximise benefits and rewards for their large-scale investments into Big Science projects. One example is procedures of fair return. Here, the total value of contracts awarded to domestic companies should, on a long-term basis, correspond to the contribution of the country to the project budget. Another example is in-kind contributions, which is the delivery of (pre-) manufactured goods, instead of direct financial investment. In the case of in-kind contributions, the money is thus spent domestically and expected to boost national economies. While the fair return policies were much used in scientific cooperation that was regarded as politically strategic, such as the ESRO (European Space Research Organisation) and ELDO (European Launcher Development Organisation) projects for the space sector in the 1960s, in-kind contributions seem to be an attractive bargaining tool nowadays in recent projects such as the European XFEL or ESS (European Spallation Source).[50]

2.2 What Role for the European Economic Union (EEC) and the European Union (EU)?

To the extent that many Big Science collaborations in Europe carry *Europe* or *European* in their name, these projects neither constitute an official body of the EEC/EU nor are formally entangled with European integration mechanisms and common policies. But, as mentioned in the introductory paragraph of this chapter, it is also true that the historical development of Big Science collaborations in Europe throughout the twentieth century and the early twenty-first century bears significant relevance to the changing patterns and dynamics of European politics and European integration.[51] The formation and development of the EEC/EU, the dynamics of European integration as well as the gradual establishment of a common EU research policy thus provide important backgrounds in this regard. They enrich scholarly understanding of the manifold entanglements and disentanglements between a common European framework and intergovernmental Big Science projects and how EEC/EU politics and policymaking mattered (or did not matter) for the founding of the ESRF and the European XFEL.

The political situation in Europe after the end of the Second World War was delicate as many European states seriously fell behind in terms of economic growth and technological progress. This particularly mattered in comparison to the United States that emerged as a global military and economic power but also as the spearhead of science and technology efforts in the post-war period. The creation of the ECSC (European Coal and Steel Community) in 1952, as well as the signing of the Treaty establishing the European Economic Community, along with the creation of Euratom (European Atomic Energy Community) in 1957, opened a window of opportunity towards collaboration. In the fields of coal, steel and nuclear power, new collaborative paths became set for unprecedented steps in political integration in Europe.

These treaties, however, had left the EEC without any (legal) competences in science, technology and research. The common portfolio of the EEC focused primarily on economic cooperation, including the creation of a customs union, a single market and eventually a monetary union.[52] Common research efforts existed only with regard to coal and nuclear energy that were key resources for economic growth and industrial development at that time.[53] Article 4 of the Euratom treaty provided several community-based competences in the field of industrial nuclear research and development. Over the last decades, JET (Joint European Torus) and ITER (International Thermonuclear Experimental Reactor) emerged as Big Science collaborations (officially termed as Joint Enterprises/Joint Undertakings[54]) under the auspices of Euratom, and through a common decision-making process prepared by the European Commission and concluded by the Council of Ministers.[55] JET, a research facility for fusion energy and plasma physics, was established in 1978 and located in Culham, United Kingdom. ITER, a similar, but much larger and expanded nuclear fusion research facility, was established in 2007, but its history dates back to the 1970s and 1980s.[56]

Early measures to improve European coordination and collaboration of research-related issues relevant to industry and economy were set up in the mid-1960s. Topical emphasis was put on societal concerns such as transport, telecommunication or meteorology.[57] These efforts were, on the one hand, triggered by the establishment of the OECD (Organisation for Economic Co-operation and Development) in 1961, which provided a platform to better coordinate formerly competing national science policies.[58] The controversy on the *technological gap* at that time referred to the leading US-American position in science and technology issues, in comparison to Europe, and a deficit in innovation capacity and technology transfer among European countries. This provided, on the other hand, an additional impetus to better pool and coordinate scientific and technological resources in Europe.[59] Two initiatives are particularly important: PREST (Politique de Recherche Scientifique et Technologique) within the EEC Medium-term Economic Policy Committee in 1965 and its successor COST (Cooperation Européenne dans le Domaine de la Science et de la Technologie). The aim of PREST was to compare the national performances in research and development and formulate measures and strategies that enable the EEC to create a coordinated common policy of science, research and technology. COST provided the first European forum for scientific and technological cooperation, and served as a coordinating hub for national research activities. Importantly, while COST was initiated within a common EEC-related context, it was eventually established as an intergovernmental forum for scientific and technological coordination in Europe.[60]

The first Framework Programme (FP) in 1984 marked a fundamental milestone for the EEC towards a more coherent research policy.[61] A topical core concern of the first FP was the effective use of Europe's capabilities and capacities in science and technology that meant, first and foremost, its exploitation in an application-oriented and industry-related context.[62] The Single European Act in 1986 provided the necessary legal competences for the establishment of a common European research policy, and aimed to increase the scientific and technological competitive standing of the industries in Europe.[63]

The EUREKA (European Research Coordination Agency) initiative was set up in 1985 to contribute to a better transnational cooperation in research and innovation. Its creation is important in two regards: First,

EUREKA, based on intergovernmental agreement (similar to COST), should be considered as a major counterweight to the FPs and similar EEC-initiated measures such as PREST.[64] Second, and related to the first point, EUREKA's status as an intergovernmental project can be regarded as a clear sign that European governments were still reluctant to hand power and control over to the EEC. The project's framework refrains the deep disfavour of European countries towards the way science, technology and research was inflexibly organised and carried out on the EEC level at that time, requiring unanimous agreements and lengthy negotiations, as the initiation of the first FP had illustrated.[65]

Following the establishment of the first FP in 1984 and the Single European Act in 1986, Pierre Papon argues that at the end of the 1980s "[t]he mechanics for the implementation of a European research policy were practically put in place."[66] Yet, at the end of the 1990s, European research policy was still reminiscent of an "uncoordinated patchwork"[67] as a combination of cross-border cooperation between the member countries, intergovernmental Big Science facilities, as well as the FPs and supranational competences in research, science and technology provided by the Single European Act and later the Treaty on European Union (*Maastricht Treaty*) of 1992. While the FPs provided a common programme-oriented and centrally managed structure to finance and support science, technology and research, efforts such as COST, EUREKA or intergovernmental collaborations in large-scale (and small-scale) research represented, quite differently, a de-centralised framework without a clear topical focus, but with contributions from a great variety of fields.[68]

Although the FPs grew over the course of the 1990s and 2000s by broadening budget and thematic scopes, this did not necessarily lead to a more coherent and comprehensive European research policy strategy. The scientific communities in Europe rather regarded EU funding as an additional and generous source of funding.[69] It was only by the early 2000s that a stepwise change, at least rhetorically, of EU policymaking in science, technology and research-related issues occurred. In January 2000, the European Commission issued a communication titled *Towards a European Research Area*.[70] Soon after, the idea of ERA (European Research Area), as a kind of internal market for research, became integrated into the Lisbon Strategy of the EU that announced to make the EU the world leading technological player. This development at the beginning of the twenty-first century also includes longer historical trajectories: It dates back to a critical period preceding the conclusion of the Lisbon Strategy that can be traced back to the early 1990s when the EU Commission identified the constant rise of unemployment within EU's member countries and the weak competitive position of the EU vis-à-vis the United States or Japan as fundamental reasons for EU's low economic performance in recent years.[71] Through the Lisbon Strategy and the vision of ERA, the EU had presented new policy directives, lifting the issues of science, technology and research onto the top of its common policy agenda. In the following years, the European Commission initiated and created several forums and mechanisms that should strengthen and improve the economic competitiveness of the EU and its member countries. These included, for instance, the creation of the ERC (European Research Council), a new centralised body for fundamental research funding but also a variety of policymaking instruments and principles (such as the Open Method of Coordination or the European Added Value). All of these tools and instruments were established to fulfil specific purposes and tasks, but perhaps, most importantly, they also testify to the fundamental and significant role that knowledge, science and technology came to play in the daily business of the EU since the late 1990s and early 2000s.[72]

The political ambitions of the European Commission for ERA also included a supportive framework, which was expected to improve coordination and management of Big Science projects. The European Commission started to use the umbrella term *Research Infrastructures* (RIs) that include national as well as collaborative efforts, single-sited and distributed projects, or single large instruments, networks or databanks in various disciplines and scientific fields (natural sciences and humanities alike) and other kinds of (large) scientific collaboration.[73] Sociologist Olof Hallonsten argued that the concept of Research Infrastructures is too broad and varied and analytically useless.[74] But the political expectations that are placed on RIs in Europe are high, namely, that these facilities should considerably contribute to the solving of urgent societal challenges, such as climate change, health or food security. The European Commission, for instance, characterises these projects as no less than "key drivers for European capacity building."[75] The ESFRI (European Strategy Forum on Research Infrastructures) forum, as a former informal round table of high political delegates from the science, technology and/or research ministries of the member countries, became an official EU strategy instrument in 2002. But it does not have decision-making power or an official mandate.[76] The launch of the first ESFRI roadmap in 2006 was characterised by EU administrators and policymakers as a systematic and coordinated approach to manage, access and realise Big Science collaborations in Europe.[77] In 2008, ERIC (European Research Infrastructure Consortium) was implemented and promoted as an innovative legal framework for collaborative Big Science projects that should facilitate the set-up of new Research Infrastructures in Europe by partly providing several of the advantages of costly and complex treaty organisations.[78]

Throughout the 2000s, the FPs remained the central funding instrument and supporting scheme of the EU for science, technology and research. However, EU funding of Big Science mainly mattered for the preparation phases and preparatory activities of these projects.[79] But the FPs also support concerns such as networking and coordination, good governance and management of Big Science facilities, as well as Intellectual Property Rights and legal frameworks.[80] Contributions to the construction phase of large-scale research projects is very limited, but additional funding is provided by the EU Structural Funds or the EIB (European Investment Bank).[81]

Reconciling from above, it can be argued that the portfolio of intergovernmental Big Science collaborations in Europe differs, on the one hand, from other large-scale research projects that constitute a common European resource (mainly under Euratom) and, on the other hand, (still) disentangles from common European research policy approaches that gradually developed, most notably since the mid-1980s. There is a first distinction to be drawn between intergovernmental Big Science projects in Europe (such as the ESRF and the European XFEL), around which this book centres, and common European Big Science projects and facilities (such as JET or ITER), that emerged under the auspices of Euratom. There is a second line to be drawn between, on the one hand, common European competences in research policy that successively developed along national policies over the last decades and that nowadays also include an emerging RI policy and, on the other hand, the set-up of intergovernmental Big Science projects and similar scientific collaborations that remain to be established *alongside* these common European principles, mechanisms and strategies.

Related to this, John Krige described the dynamics of intergovernmental scientific collaboration in Europe as an alternative to a path already "opened up by the restructuring of the European politico-economic space."[82] Pierre Papon adds that the overall European integration "has walked on two legs by evolving according to two different modes."[83] First, the implementation of common policies, for example agriculture, trade or money. Second, the ad-hoc creation of intergovernmental cooperation, merely in the fields of science, technology and research, on the other hand. To put it differently, while economic and political European integration has thoroughly deepened since the late 1950s by establishing integrated European spaces such as the customs union, the single market, the Euro zone and the EU per se, such an integrated space does not yet fully exist for research, science and technology. Whether ERA can become the kind of internal market for research the European Commission originally envisaged remains a topical issue of current scholarly research.[84]

The historical development of Big Science collaborations alongside the development of common European policies for science, technology and research can partly be explained through the observation that concerns in science, technology and research in Europe often touched upon national strategic interests of individual countries. Integration of these topics into a common European frame was, moreover, complicated by the divergent structures and legal frameworks of the national science systems and their institutions. Lifting intergovernmental collaboration onto a common, integrated European level would have meant that funding, decision-making processes and organisation of these projects were carried out within and based on the work of common European bodies, such as the European Commission and/or the Council of Ministers, which required to hand over power and control of these issues.[85] Common ground for action could, as argued above, mainly be found in the fields of monetary, trade and agricultural policies, which were among the first politically integrated fields. But how to explain the gradual establishment of a common RI policy? It seems as if these dynamics link to a change in the understanding and conception of integration. Scholarly research revealed that in the early years of the EEC, successful integration was perceived as a shift of sovereignty from national to supranational frameworks. This is also supported by John Krige arguing that:

> collaboration in a European scientific organization always involves a loss of, or at least a dilution of, national sovereignty. This loss is accepted, but not taken for granted. Its scope is limited, carefully monitored and constantly re-evaluated.[86]

But since the 2000s, it can be observed that (successful) integration does no longer necessarily involve a shift of sovereignty, but grounds, first and foremost, on improved coordination. It may hence be argued that to the extent that coordination also became increasingly perceived as a means to improve and strengthen national efforts, research policy competences of the EU could be deepened and RI policy could emerge.[87]

Notes

1. C. Defrance, "France-Allemagne: Une Coopération Scientifique 'Privilégiée' en Europe, de l'Immédiat Après-Guerre au Milieu des Années 1980?" In *La Guerre Froide et l'Internationalisation des Sciences: Acteurs, Réseaux et Institutions*, eds. C. Defrance and A. Kwaschik (Paris: CNRS, 2016), 182. Translated from French.
2. In the following, the term *common* refers to different kinds of political integration in the EEC/EU (most importantly, differentiated and uniform integration) by which at least some of the national competences where transferred to the supranational bodies and institutions of the EEC/EU.
3. See, for example, P. Papon, "European Scientific Cooperation and Research Infrastructures: Past Tendencies and Future Prospects." *Minerva* 42, no. 1 (2004); P. Papon, "Intergovernmental Cooperation in the Making of European Research." In *European Science and Technology Policy: Towards Integration or Fragmentation?* eds. H. Delanghe, U. Muldur, and L. Soete (Cheltenham: Edward Elgar, 2009); P. Papon, "L'Espace Européen de la Recherche (1960–1985): Entre Science et Politique." In *La Construction d'un Espace Scientifique Commun? La France, la RFA et l'Europe après le "Choc du Spoutnik"*, eds. C. Defrance and U. Pfeil (Bruxelles, New York: P.I.E. Peter Lang, 2012); J. Krige, "The Politics of European Scientific Collaboration." In *Companion to Science in the Twentieth Century,* eds. J. Krige and D. Pestre (London: Routledge, 2003); O. Hallonsten, "The Politics of European Collaboration in Big Science." In *The Global Politics of Science and Technology*, Vol. 2, eds. M. Mayer, M. Carpes, and R. Knoblich (Berlin, Heidelberg: Springer, 2014); O. Hallonsten, "Contextualising the European Spallation Source: What We Can Learn from the History, Politics, and Sociology of Big Science." In *In Pursuit of a Promise: Perspectives on the Political Process to Establish the European Spallation Source (ESS) in Lund, Sweden*, ed. O. Hallonsten (Lund: Arkiv Academic Press, 2012).
4. O. Hallonsten, "Research Infrastructures in Europe: The Hype and the Field." *European Review* 28, no. 2 (2020); T. Stahlecker and H. Kroll, *Policies to Build Research Infrastructures in Europe: Following Traditions or Building New Momentum?* (Karlsruhe: Fraunhofer ISI, 2013).
5. Tensions of Europe (ToE) is an international scholarly network founded in 1999 that researches and promotes the role and significance of technology

in European history. See the website of the ToE network: https://www.tensionsofeurope.eu

6. The focus is put on technology and infrastructures in a general sense. Large/big technologies constitute one perspective among many others.
7. See, for example, T. Misa and J. Schot, "Inventing Europe: Technology and the Hidden Integration of Europe." *History and Technology* 21, no. 1 (2005); H. Trischler and H. Weinberger, "Engineering Europe: Big Technologies and Military Systems in the Making of 20th Century Europe." *History and Technology* 21, no. 1 (2005); F. Schipper and J. Schot "Infrastructural Europeanism, or the Project of Building Europe on Infrastructures: An Introduction." *History and Technology* 27, no. 3 (2011); A. Badenoch and A. Fickers, *Materializing Europe: Transnational Infrastructures and the Project of Europe* (New York, NY: Palgrave Macmillan, 2010); E. van der Vleuten, "Large Technical Systems." In *A Companion to the Philosophy of Technology*, eds. by J. Friis, S. Pedersen, and V. Hendricks (Chichester: Wiley-Blackwell, 2009).
8. See, for example, Misa and Schot, "Inventing Europe", 2, 3, 7.
9. K. K. Patel, "Provincialising European Union: Co-Operation and Integration in Europe in a Historical Perspective." *Contemporary European History* 22, no. 04 (2013), 650.
10. Patel, "Provincialising European Union" (2013), 651.
11. H. Trischler, "Physics and Politics. Research and Research Support in Twentieth Century Germany in International Perspective – An Introduction." In *Physics and Politics: Research and Research Support in Twentieth Century Germany in International Perspective*, eds. H. Trischler and M. Walker (Stuttgart: Franz Steiner Verlag, 2010), 10. See also, M. Ash, "Wissenschaft und Politik als Ressourcen füreinander." In *Wissenschaften und Wissenschaftspolitik: Bestandsaufnahmen zu Formationen, Brüchen und Kontinuitäten im Deutschland des 20. Jahrhunderts*, eds. R. vom Bruch and B. Kaderas (Stuttgart: Franz Steiner, 2002).
12. See, for example, M. Ash, "Wissenschaft und Politik: Eine Beziehungsgeschichte im 20. Jahrhundert." *Archiv für Sozialgeschichte* 50 (2010), 15–18. See also, P. Weingart, "Verwissenschaftlichung der Gesellschaft – Politisierung der Wissenschaft." *Zeitschrift für Soziologie* 12 (1983); L. Raphael, "Die Verwissenschaftlichung des Sozialen als methodische und konzeptionelle Herausforderung für eine Sozialgeschichte des 20. Jahrhunderts." *Geschichte und Gesellschaft* 22, no. 2 (1996); P. Weingart, "Scientific Expertise and Political Accountability: Paradoxes of Science in Politics." *Science and Public Policy* 26, no. 3 (1999). For a similar perspective with regard to Big Science, see, for example, G. Ritter,

Großforschung und Staat in Deutschland: Ein historischer Überblick (München: Beck, 1992); O. Hallonsten, Big *Science Transformed: Science, Politics and Organization in Europe and the United States* (Cham: Palgrave Macmillan, 2016), 213.
13. Trischler, "Physics and Politics", 10.
14. See, for example, Papon "European Scientific Cooperation"; Papon, "Intergovernmental Cooperation"; Krige, "The Politics of European Scientific Collaboration"; Hallonsten, "The Politics of European Collaboration"; Hallonsten *Big Science Transformed*; J. Krige and D. Pestre (eds.), *Companion to Science in the Twentieth Century* (London: Routledge, 2003).
15. ESF (ed.), *ESF EUROCORES Programme "Inventing Europe – Technology and the Making of Europe, 1850 to the Present": Final Report* (Strasbourg: European Science Foundation, 2011), 3.
16. ESF, *EUROCORES*, 3.
17. J. Capshew and K. Rader, "Big Science: Price to the Present." *Osiris* 7, Science after '40 (1992), 8.
18. R. Li-Hua, "Definitions of Technology." In *A Companion to the Philosophy of Technology*, eds. J. Friis, S. Pedersen and V. Hendricks (Chichester: Wiley-Blackwell, 2009), 19.
19. See Li-Hua, "Definitions of Technology", 19.
20. S. Jasanoff, "Ordering Knowledge, Ordering Society." In *States of Knowledge: The Co-production of Science and the Social Order*, ed. S. Jasanoff (London: Routledge, 2006), 16.
21. See, for example, P. Castelnovo, M. et al., "The Economic Impact of Technological Procurement for Large-Scale Research Infrastructures: Evidence from the Large Hadron Collider at CERN." *Research Policy* 47, no. 9 (2018); Technopolis Group, *Big Science and Innovation* (Brighton: Technopolis Group, 2013).
22. ESF, *EUROCORES*, 3.
23. E. van der Vleuten, "Toward a Transnational History of Technology: Meanings, Promises, Pitfalls." *Technology and Culture* 49, no. 4 (2008), 976.
24. Schipper and Schot, "Infrastructural Europeanism", 252.
25. For example, European Spallation Source, European Southern Observatory, European Molecular Biology Laboratory.
26. van der Vleuten, "Toward a Transnational History", 985.
27. J. Krige, "Embedding the National in the Global." In *Science and Technology in the Global Cold War*, eds. N. Oreskes and J. Krige (Cambridge, Mass.: MIT Press, 2014), 228.
28. Krige, "Embedding the National", 228.

29. Krige, "Embedding the National", 228.
30. Krige, "Embedding the National", 245.
31. Krige, "The Politics of European Scientific Collaboration", 900.
32. Krige, "Embedding the National", 245.
33. See van der Vleuten, "Toward a Transnational History".
34. See, for example, Krige, "The Politics of European Scientific Collaboration", Hallonsten, "The Politics of European Collaboration", Hallonsten, *Big Science Transformed.*
35. See, for example, Defrance, "France-Allemagne", 181.
36. See, for example, Krige, "The Politics of European Scientific Collaboration", 908; Papon, "Intergovernmental Cooperation", 39; Defrance, "France-Allemagne", 181.
37. See, for example, K. Cramer, "The Role of European Big Science in the (Geo)Political Challenges of the Twentieth and Twenty-First Centuries." In *Big Science and Research Infrastructures in Europe*, eds. K. Cramer and O. Hallonsten (Cheltenham: Edward Elgar, 2020).
38. J. Krige, "Preface." In *Choosing Big Technologies*, ed. J. Krige (Chur: Harwood Academic Publishers, 1993), viii.
39. See J. Krige, "Historical Synthesis." In *History of European Scientific and Technological Cooperation*, eds. J. Krige and L. Guzzetti (Luxembourg: European Communities, 1997), 442–443.
40. See, for example, Hallonsten, "The Politics of European Collaboration"; Krige, "The Politics of European Scientific Collaboration"; Hallonsten, "Continuity and Change".
41. See Cramer, "The Role of European Big Science".
42. See, for example, Papon, "Intergovernmental Cooperation", 36; L. Guzzetti, *A Brief History of European Union Research Policy* (Luxembourg: European Communities, 1995), 15, 29, 42, 47; Papon, "European Scientific Cooperation", 68. See also, J.-L. Roland, "COST: An Unexpected Successful Cooperation." In *History of European Scientific and Technological Cooperation*, eds. J. Krige and L. Guzzetti (Luxembourg: European Communities, 1997), 356; J. Peterson, "Eureka: A Historical Perspective." In *History of European Scientific and Technological Cooperation*, eds. J. Krige and L. Guzzetti (Luxembourg: European Communities, 1997). Roland and Peterson apply the "variable geometry" and "Europe à la carte" vocabulary to the COST and EUREKA projects.
43. See, for example, M. Goldsmith, "Variable Geometry, Multilevel Governance: European Integration and Subnational Government in the New Millennium." In *The Politics of Europeanization*, eds. K. Featherstone and C. Radaelli (Oxford University Press, 2003); F. Tassinari, *Variable Geometries. Mapping Ideas, Institutions and Power in the Wider Europe*, CEPS Working Document No. 254 (2006).

44. See, for example, R. Williams, "Choosing Big Technologies: The Core Issues." In *Choosing Big Technologies*, ed. J. Krige (Chur: Harwood Academic Publishers, 1993), 224.
45. See, for example, A. Fazekas, A. "Big Science in a Small Country." *Science*, 20 January 2006; S. Widmalm, "Big Science in a Small Country: Sweden and CERN II." In *Center on the Periphery: Historical Aspects of 20th-Century Swedish Physics*, ed. S. Lindqvist (Canton: Warson Publishing International, 1993); M. Heymann and J. Martin-Nielsen, "Introduction: Perspectives on Cold War Science in Small European States." *Centaurus* 55 (2013).
46. See, for example, O. Hallonsten, "Is There an 'Iron Law' of Big Science?" In *Big Science and Research Infrastructures in Europe*, eds. K. Cramer and O. Hallonsten (Cheltenham: Edward Elgar, 2020); O. Hallonsten, "Myths and Realities of the ESS project: A Systematic Scrutiny of Readily Accepted 'Truths'." In *Legitimizing ESS: Big Science as a Collaboration Across Boundaries*, eds. T. Kaiserfeld, and T. O'Dell (Lund: Nordic Academic Press, 2013), 56ff; M. Riordan, L. Hoddeson, and A. Kolb, *Tunnel Visions: The Rise and Fall of the Superconducting Super Collider* (Chicago: University of Chicago Press, 2015), 98ff; L. Hoddeson, A. Kolb, and C. Westfall, *Fermilab: Physics, the Frontier, and Megascience* (Chicago: University of Chicago Press, 2008), 70ff.
47. See O. Hallonsten, "Continuity and Change in the Politics of European Scientific Collaboration." *Journal of Contemporary European Research* 8, no. 3 (2012), 304.
48. See Krige, "The Politics of European Scientific Collaboration", 906.
49. Hallonsten, "Continuity and Change", 311.
50. See, for example, Hallonsten, *Big Science Transformed*, 93, 120.
51. See, for example, Trischler and Weinberger, "Engineering Europe", 75.
52. See P. Tindemans, "Post-War Research, Education and Innovation Policy-Making in Europe." In *European Science and Technology Policy: Towards Integration or Fragmentation?* eds. H. Delanghe, U. Muldur, and L. Soete (Cheltenham: Edward Elgar, 2009) 13.
53. See Guzzetti, *A Brief History*, 4, 11.
54. At the time of ITER's establishment, Joint Enterprises were already renamed Joint Undertakings. See *Treaty Establishing the European Atomic Energy Community*, Chapter V.
55. See *Treaty Establishing the European Atomic Energy Community*, Chapter V, Art. 45–47.
56. See McCray, "'Globalization with Hardware': ITER's Fusion of Technology, Policy, and Politics." *History and Technology* 26, no. 4 (2010).
57. See Tindemans, "Post-War Research", 14; Guzzetti, *A Brief History*, 39–44.

58. See OECD, *The Measurement of Scientific and Technical Activities: Proposed Standard Practice for Surveys of Research and Experimental Development* (Paris: OECD Publishing, 1962).
59. See, for example, J.-J. Servan-Schreiber, *Le Défi Américain* (Paris: Denoël, 1967); OECD, *Gaps in Technology. Analytical Report: Comparison between Member Countries in Education, Research Development, Technological Innovation, International Economic Exchange* (Paris: OECD Publishing, 1970).
60. See Guzzetti, *A Brief History*, 39–42.
61. See Guzzetti, *A Brief History*, 83.
62. See Guzzetti, *A Brief History*, 83–86.
63. See Tindemans, "Post-War Research", 17.
64. See Peterson, "Eureka", 326–327.
65. See Peterson, "Eureka", 336.
66. Papon, "L'Espace Européen", 49. Translated from French.
67. H. Delanghe, U. Muldur, and L. Soete, "Conclusion and Perspectives." In *European Science and Technology Policy: Towards Integration or Fragmentation?* Eds. H. Delanghe, U. Muldur, and L. Soete (Cheltenham: Edward Elgar, 2009), 354.
68. See P. Tindemans, "Post-War Research", 15. Similar: Hallonsten, "Continuity and Change", 302.
69. See A. de Elera, "The European Research Area: On the Way Towards a European Scientific Community?" *European Law* 12, no. 5 (2006), 562.
70. European Commission. *Communication from the Commission to the Council, the European Parliament, the Economic and Social Committee and the Committee of the Regions – Towards a European Research Area* (Brussels, 2000).
71. See European Commission, *Growth, Competitiveness, Employment. The Challenges and Ways Forward into the 21st Century*, White paper (Luxemburg: Office for Official Publications of the European Commission, 1993).
72. See, for example, de Elera, "The European Research Area"; I. Ulnicane, "Research and Innovation as Sources of Renewed Growth? EU Policy Responses to the Crisis." *Journal of European Integration* 38, no. 3 (2016); I. Ulnicane, "Broadening Aims and Building Support in Science, Technology and Innovation Policy: The Case of the European Research Area." *Journal of Contemporary European Research* 11, no. 1 (2015).
73. European Commission, *Developing World-class Research Infrastructures for the European Research area (ERA): Report of the ERA Expert Group* (Luxembourg: Office for Official Publications of the European Communities, 2008), 14 fn. 1.
74. See Hallonsten, "The Hype and the Field".

75. European Commission, "Developing World-Class Research Infrastructures", 15.
76. I. Bolliger and A. Griffiths, "The Introduction of ESFRI and the Rise of National Research Infrastructure Roadmaps in Europe." In *Big Science and Research Infrastructures in Europe*, eds. K. Cramer and O. Hallonsten (Cheltenham: Edward Elgar, 2020).
77. See Bolliger and Griffiths, "The Introduction of ESFRI".
78. See M. Moskovko, A. Astvaldsson, and O. Hallonsten, "Who is ERIC? The Politics and Jurisprudence of a New Governance Tool for Collaborative European Research Infrastructures", *Journal of Contemporary European Research* 15, no. 3; Moskovko, M. "Intensified Role of the EU? European Research Infrastructure Consortium (ERIC) as a Legal Framework for Contemporary Multinational Research Collaboration." In *Big Science and Research Infrastructures in Europe*, eds. K. Cramer and O. Hallonsten (Cheltenham: Edward Elgar, 2020).
79. The seventh FP of the EU funded the Pre-XFEL project as a preparatory effort for the European XFEL project, and the ESRFUP project as a preparatory effort for the upgrade programme of the ESRF.
80. Examples include RAMIRI (Realising and Managing International Research Infrastructures) or RItrain (Research Infrastructure Training Programme).
81. H. Pero, "Research Infrastructures of Pan-European Interest: The EU and Global Issues." *Nuclear Instruments and Methods in Physics Research Section A* 626–627 (2011).
82. Krige, "The Politics of European Scientific Collaboration", 898.
83. Papon, "Intergovernmental Cooperation", 39.
84. See, for example, T. Luukkonen, "European Research Area: An Evolving Policy Agenda." In *Towards European Science. Dynamics and Policy of an Evolving European Research Space*, eds. L. Wedlin and M. Nedeva (Cheltenham: Edgward Elgar, 2015); T. Banchoff, "Political Dynamics of the ERA." In *Changing Governance of Research and Technology Policy: The European Research Area*, eds. J. Edler, S. Kuhlmann, and M. Behrens (Cheltenham: Edward Elgar, 2003); N. Nedeva, "Between the Global and the National: Organising European Science." *Research Policy* 42, no. 1 (2013).
85. See, for example, Krige, "The Politics of European Scientific Collaboration", 900.
86. Krige, "The Politics of European Scientific Collaboration", 900.
87. Banchoff, "Political Dynamics", 83–84.

Bibliography

Ash, M. "Wissenschaft und Politik als Ressourcen füreinander." In *Wissenschaften und Wissenschaftspolitik: Bestandsaufnahmen zu Formationen, Brüchen und Kontinuitäten im Deutschland des 20. Jahrhunderts*, edited by R. vom Bruch and B. Kaderas, 32–51. Stuttgart: Franz Steiner, 2002.

Ash, M. "Wissenschaft und Politik: Eine Beziehungsgeschichte im 20. Jahrhundert." *Archiv für Sozialgeschichte* 50 (2010): 11–46.

Badenoch, A. and Fickers, A. *Materializing Europe: Transnational Infrastructures and the Project of Europe*. New York, NY: Palgrave Macmillan, 2010.

Banchoff, T. "Political Dynamics of the ERA." In *Changing Governance of Research and Technology Policy: The European Research Area*, edited by J. Edler, S. Kuhlmann, and M. Behrens, 81–97. Cheltenham: Edward Elgar, 2003.

Bolliger, I. and Griffiths, A. "The Introduction of ESFRI and the Rise of National Research Infrastructure Roadmaps in Europe." In *Big Science and Research Infrastructures in Europe*, edited by K. Cramer and O. Hallonsten, 101–127. Cheltenham: Edward Elgar, 2020.

Capshew, J. and Rader, K. "Big Science: Price to the Present." *Osiris* 7, Science after '40 (1992): 2–25.

Castelnovo, P., Florio, M., Forte, S., Rossi, L., and Sirtori, E. "The Economic Impact of Technological Procurement for Large-Scale Research Infrastructures: Evidence from the Large Hadron Collider at CERN." *Research Policy* 47, no. 9 (2018): 1853–1867.

Cramer, K. "The Role of European Big Science in the (Geo)Political Challenges of the Twentieth and Twenty-First Centuries." In *Big Science and Research Infrastructures in Europe*, edited by K. Cramer and O. Hallonsten, 56–75. Cheltenham: Edward Elgar, 2020.

de Elera, A. "The European Research Area: On the Way Towards a European Scientific Community?" *European Law* 12, no. 5 (2006): 559–574.

Defrance, C. "France-Allemagne: Une Coopération Scientifique 'Privilégiée' en Europe, de l'Immédiat Après-Guerre au Milieu des Années 1980?" In *La Guerre Froide et l'Internationalisation des Sciences: Acteurs, Réseaux et Institutions*, edited by C. Defrance and A. Kwaschik, 169–186. Paris: CNRS, 2016.

Delanghe, H., Muldur, U., and Soete L. "Conclusion and Perspectives." In *European Science and Technology Policy: Towards Integration or Fragmentation?* edited by H. Delanghe, U. Muldur, and L. Soete, 353–56. Cheltenham: Edward Elgar, 2009.

ESF, ed. *ESF EUROCORES Programme "Inventing Europe – Technology and the Making of Europe, 1850 to the Present": Final Report*. Strasbourg: European Science Foundation, 2011.

European Commission. *Developing World-class Research Infrastructures for the European Research area (ERA): Report of the ERA Expert Group*. Luxembourg: Office for Official Publications of the European Communities, 2008.
European Commission. *Growth, Competitiveness, Employment. The Challenges and Ways Forward into the 21st Century*, White paper. Luxemburg: Office for Official Publications of the European Commission, 1993.
European Commission. *Communication from the Commission to the Council, the European Parliament, the Economic and Social Committee and the Committee of the Regions – Towards a European Research Area*. Brussels, 2000.
Fazekas, A. "Big Science in a Small Country." *Science*, 20 January 2006. Online available at: https://www.sciencemag.org/careers/2006/01/big-science-small-country (last accessed 10 December 2019).
Goldsmith, M. "Variable Geometry, Multilevel Governance: European Integration and Subnational Government in the New Millennium." In *The Politics of Europeanization*, edited by K. Featherstone and C. Radaelli, 112–133. Oxford University Press, 2003.
Guzzetti, L. *A Brief History of European Union Research Policy*. Luxembourg: European Communities, 1995.
Hacking, I. "The Self-Vindication of the Laboratory Sciences." In *Science as Practice and Culture*, edited by A. Pickering, 29–64. Chicago: University of Chicago Press, 1992.
Hallonsten, O. "Is There an 'Iron Law' of Big Science?" In *Big Science and Research Infrastructures in Europe*, edited by K. Cramer and O. Hallonsten, 217–231. Cheltenham: Edward Elgar, 2020a.
Hallonsten, O. "Research Infrastructures in Europe: The Hype and the Field." *European Review* 28, no. 2 (2020b), 1–19.
Hallonsten, O. *Big Science Transformed: Science, Politics and Organization in Europe and the United States*. Cham: Palgrave Macmillan, 2016.
Hallonsten, O. "The Politics of European Collaboration in Big Science." In *The Global Politics of Science and Technology*, Vol. 2, edited by M. Mayer, M. Carpes, and R. Knoblich, 31–46. Berlin, Heidelberg: Springer, 2014.
Hallonsten, O. "Myths and Realities of the ESS project: A Systematic Scrutiny of Readily Accepted 'Truths'." In *Legitimizing ESS: Big Science as a Collaboration Across Boundaries*, edited by T. Kaiserfeld, and T. O'Dell, 43–66. Lund: Nordic Academic Press, 2013.
Hallonsten, O. "Contextualising the European Spallation Source: What We Can Learn from the History, Politics, and Sociology of Big Science." In *In Pursuit of a Promise: Perspectives on the Political Process to Establish the European Spallation Source (ESS) in Lund, Sweden*. Edited by O. Hallonsten, 81–107. Lund: Arkiv Academic Press, 2012a.

Hallonsten, O. "Continuity and Change in the Politics of European Scientific Collaboration." *Journal of Contemporary European Research* 8, no. 3 (2012b): 300–319.

Heymann, M. and Martin-Nielsen, J. "Introduction: Perspectives on Cold War Science in Small European States." *Centaurus* 55 (2013): 221–242.

Hoddeson, L., Kolb, A. and Westfall, C. *Fermilab: Physics, the Frontier, and Megascience*. Chicago: University of Chicago Press, 2008.

Jasanoff, S. "Ordering Knowledge, Ordering Society." In *States of Knowledge: The Co-production of Science and the Social Order*, edited by S. Jasanoff, 13–45. London: Routledge, 2006.

Krige, J. "Embedding the National in the Global." In *Science and Technology in the Global Cold War*, edited by N. Oreskes and J. Krige, 227–250. Cambridge, MA: MIT Press, 2014.

Krige, J. "The Politics of European Scientific Collaboration." In *Companion to Science in the Twentieth Century*, edited by J. Krige and D. Pestre, 897–918. London: Routledge, 2003.

Krige, J. "Historical Synthesis." In *History of European Scientific and Technological Cooperation*, edited by J. Krige and L. Guzzetti, 439–444. Luxembourg: European Communities, 1997.

Krige, J. "Preface." In *Choosing Big Technologies*, edited by J. Krige, vii-xii. Chur: Harwood Academic Publishers, 1993.

Krige, J. and D. Pestre (eds.). *Companion to Science in the Twentieth Century* (London: Routledge, 2003).

Li-Hua, R. "Definitions of Technology." In *A Companion to the Philosophy of Technology*, edited by J. Friis, S. Pedersen, and V. Hendricks, 18–22. Chichester: Wiley-Blackwell, 2009.

Luukkonen, T. "European Research Area: An Evolving Policy Agenda." In *Towards European Science. Dynamics and Policy of an Evolving European Research Space*, edited by L. Wedlin and M. Nedeva, 37–60. Cheltenham: Edgward Elgar, 2015.

McCray, W. "'Globalization with Hardware': ITER's Fusion of Technology, Policy, and Politics." *History and Technology* 26, no. 4 (2010): 283–312.

Misa T., and Schot, J. "Inventing Europe: Technology and the Hidden Integration of Europe." *History and Technology* 21, no. 1 (2005): 1–19.

Moskovko, M. "Intensified Role of the EU? European Research Infrastructure Consortium (ERIC) as a Legal Framework for Contemporary Multinational Research Collaboration. In *Big Science and Research Infrastructures in Europe*, edited by K. Cramer and O. Hallonsten, 128–156. Cheltenham: Edward Elgar, 2020.

Moskovko, M., Astvaldsson, A., and Hallonsten, O. "Who is ERIC? The Politics and Jurisprudence of a New Governance Tool for Collaborative European

Research Infrastructures." *Journal of Contemporary European Research* 15, no. 3 (2019): 249–268.

Nedeva, N. "Between the Global and the National: Organising European Science." *Research Policy* 42, no. 1 (2013): 220–230.

OECD. *The Research System: Comparative Survey of the Organization and Financing of Fundamental Research*, Volume I: France, Germany, United Kingdom; Volume II: Belgium, Netherlands, Sweden, Switzerland; Volume III: Canada, United States, General Conclusions. Paris: OECD Publishing, 1972–1974.

OECD. *Gaps in Technology. Analytical Report: Comparison between Member Countries in Education, Research Development, Technological Innovation, International Economic Exchange*. Paris: OECD Publishing, 1970.

OECD. *The Measurement of Scientific and Technical Activities: Proposed Standard Practice for Surveys of Research and Experimental Development.* Paris: OECD Publishing, 1962.

Papon, P. "L'Espace Européen de la Recherche (1960–1985): Entre Science et Politique." In *La Construction d'un Espace Scientifique Commun? La France, la RFA et l'Europe après le "Choc du Spoutnik"*, edited by C. Defrance and U. Pfeil, 37–54. Bruxelles, New York: P.I.E. Peter Lang, 2012.

Papon, P. "Intergovernmental Cooperation in the Making of European Research." In *European Science and Technology Policy: Towards Integration or Fragmentation?* edited by H. Delanghe, U. Muldur, and L. Soete, 24–43. Cheltenham: Edward Elgar, 2009.

Papon, P. "European Scientific Cooperation and Research Infrastructures: Past Tendencies and Future Prospects." *Minerva* 42, no. 1 (2004): 61–76.

Patel, K. K. "Provincialising European Union: Co-Operation and Integration in Europe in a Historical Perspective." *Contemporary European History* 22, no. 4 (2013): 649–673.

Pero, H. "Research Infrastructures of Pan-European Interest: The EU and Global Issues." *Nuclear Instruments and Methods in Physics Research Section A* 626–627 (2011): 69–71.

Peterson, J. "Eureka: A Historical Perspective." In *History of European Scientific and Technological Cooperation*, edited by J. Krige and L. Guzzetti, 323–345. Luxembourg: European Communities, 1997.

Raphael, L. "Die Verwissenschaftlichung des Sozialen als methodische und konzeptionelle Herausforderung für eine Sozialgeschichte des 20. Jahrhunderts." *Geschichte und Gesellschaft* 22, no. 2 (1996): 165–193.

Riordan, M., Hoddeson, L., and Kolb, A. *Tunnel Visions: The Rise and Fall of the Superconducting Super Collider.* Chicago: University of Chicago Press, 2015.

Ritter, G. *Großforschung und Staat in Deutschland: Ein historischer Überblick.* München: Beck, 1992.

Roland, J.-L. "COST: An Unexpected Successful Cooperation." In *History of European Scientific and Technological Cooperation*, eds. J. Krige and L. Guzzetti, 355–368. Luxembourg: European Communities, 1997.

Schipper, F., and Schot, J. "Infrastructural Europeanism, or the Project of Building Europe on Infrastructures: An Introduction." *History and Technology* 27, no. 3 (2011): 245–264.

Servan-Schreiber, J.-J. *Le Défi Américain*. Paris: Denoël, 1967.

Stahlecker, T. and Kroll, H. *Policies to Build Research Infrastructures in Europe: Following Traditions or Building New Momentum?* Karlsruhe: Fraunhofer ISI, 2013.

Tassinari, F. *Variable Geometries. Mapping Ideas, Institutions and Power in the Wider Europe*, CEPS Working Document No. 254 (2006).

Technopolis Group. *Big Science and Innovation*. Brighton: Technopolis Group, 2013.

Tindemans, P. "Post-War Research, Education and Innovation Policy-Making in Europe." In *European Science and Technology Policy: Towards Integration or Fragmentation?* edited by H. Delanghe, U. Muldur, and L. Soete, 9–18. Cheltenham: Edward Elgar, 2009.

Trischler, H. "Physics and Politics. Research and Research Support in Twentieth Century Germany in International Perspective – An Introduction." In *Physics and Politics: Research and Research Support in Twentieth Century Germany in International Perspective*, edited by H. Trischler and M. Walker, 9–18. Stuttgart: Franz Steiner Verlag, 2010.

Trischler, H., and Weinberger, H. "Engineering Europe: Big Technologies and Military Systems in the Making of 20th Century Europe." *History and Technology* 21, no. 1 (2005): 49–83.

Ulnicane, I. "Research and Innovation as Sources of Renewed Growth? EU Policy Responses to the Crisis." *Journal of European Integration* 38, no. 3 (2016): 327–341.

Ulnicane, I. "Broadening Aims and Building Support in Science, Technology and Innovation Policy: The Case of the European Research Area." *Journal of Contemporary European Research* 11, no. 1 (2015): 31–49.

van der Vleuten, E. "Large Technical Systems." In *A Companion to the Philosophy of Technology*, edited by J. Friis, S. Pedersen, and V. Hendricks, 218–222. Chichester: Wiley-Blackwell, 2009.

van der Vleuten, E. "Toward a Transnational History of Technology: Meanings, Promises, Pitfalls." *Technology and Culture* 49, no. 4 (2008): 974–994.

Weingart, P. "Scientific Expertise and Political Accountability: Paradoxes of Science in Politics." *Science and Public Policy* 26, no. 3 (1999): 151–161.

Weingart, P. "Verwissenschaftlichung der Gesellschaft – Politisierung der Wissenschaft." *Zeitschrift für Soziologie* 12 (1983): 225–241.

Widmalm, S. "Big Science in a Small Country: Sweden and CERN II." In *Center on the Periphery: Historical Aspects of 20th-Century Swedish Physics*, edited by S. Lindqvist, 107–140. Canton: Warson Publishing International, 1993.

Williams, R. "Choosing Big Technologies: The Core Issues." In *Choosing Big Technologies*, edited by J. Krige, 223–234. Chur: Harwood Academic Publishers, 1993.

CHAPTER 3

History and Science of Research with Synchrotron Radiation

Synchrotron radiation is electromagnetic radiation, commonly known as light, which is emitted on accelerators (linear or circular) when particles are accelerated close to the speed of light and when the particle beam is redirected and disturbed through a magnetic field. The first visual observation of synchrotron radiation occurred in 1947 at the General Electric 70 MeV synchrotron in Schenectady, New York in the United States.[1] But early theoretical considerations on synchrotron radiation date back to the late nineteenth century.[2] The name synchrotron radiation relates to the fact that it was first observed at a synchrotron. But storage rings or any kind of bending magnets can, for instance, also be used as a source for this specific radiation. Research with synchrotron radiation can also be grouped under the umbrella term of photon science. In line with the fundamental principle of the wave-particle duality, light can be interpreted as both wave and particle. The particles of light are called photons, which gave this field of research its name.[3]

Synchrotron radiation is used by very different disciplines and scientific fields for a broad array of experimental purposes and applications. For instance, material sciences use synchrotron radiation to study the properties of interfaces or surfaces, or the behaviour of materials such as semiconductors, glasses and fibres.[4] Chemists or biologists investigate the structures of single proteins or molecules at synchrotron radiation sources.[5] Medical research uses synchrotron radiation to develop and design new drugs and vaccines.[6] Earth sciences study materials from the surface and

K. C. Cramer, *A Political History of Big Science*,
Palgrave Studies in the History of Science and Technology,
https://doi.org/10.1007/978-3-030-50049-8_3

the core of the earth with the help of synchrotron radiation to understand volcanic eruptions as well as plate tectonics. The exploration of nanostructures under extreme external conditions such as high temperature, ultra-high vacuum or high pressure is of common interest for a broad variety of scientific fields, such as physics, chemistry, biology, life sciences, materials sciences and geology.[7] Art history or archaeology investigates paintings, textiles or ceramics at synchrotron radiation sources revealing important molecular information on pigments, paint layers or the type of fibres.[8] Diffraction, scattering and spectroscopy are three widely used techniques among several others in research with synchrotron radiation. They are, however, only umbrella terms for a great number of specialised and highly sophisticated variations of techniques.[9] Current research opportunities at free-electron lasers additionally open the possibility to study non-crystalline samples and to investigate the change in structure of very small building blocks.[10]

Most synchrotron radiation sources or free-electron lasers (including the ESRF and the European XFEL projects) accelerate electrons or positrons. Although it is possible to make use of other particles, only these two kinds of particles have a velocity close enough to the speed of light that can produce such intensive radiation that serves today's experimental needs, most notably the investigation of living matter or materials.[11] Synchrotron radiation covers a broad range of wavelengths in the electromagnetic spectrum: from IR (infrared), to UV (ultraviolet), VUV (vacuum-ultraviolet), XUV (extreme ultraviolet) to the soft and hard X-ray regime, to name the most common spectral ranges.[12] The hard X-ray range is of particular interest because its wavelength of approximately 1 Ångström, that is 0.1 nm, equals the atomic dimension. This is crucial for the study of matter and materials at the level of atoms that are less than a nanometre in size. The X-ray wavelength regime (together with the improvement of related parameters) became a kind of gold standard for cutting-edge synchrotron radiation sources and free-electron lasers.[13] The ESRF and the European XFEL both operate in the hard X-ray regime.

The history of research with synchrotron radiation cannot be told without reference to the history of particle physics in the sense that synchrotron radiation emerged as an unwanted by-product of particle physics experiments in the 1940s and 1950s. Accelerators for particle physics are purpose-built machines, and most projects mainly focus on the investigation of specific elementary particles or forces. In other words, most particle physics projects were *mission-oriented*.[14] Researchers with synchrotron radiation were, in contrast, interested in the multidisciplinary use and exploitation of the radiation emitted at particle accelerators (which particle physicists characterised as an unwanted by-product) to study materials, surfaces, liquids and living matter. In other words, research with synchrotron radiation can be characterised as *open-end*.[15] Scientific progress in particle physics required the construction of ever-larger accelerators that can accelerate particles to ever-higher energies. However, synchrotron radiation occurs as the inevitable loss of energy when magnets bend charged and accelerated particles. While the loss of energy and the amount of emitted radiation is negligible in small accelerators, the larger the accelerators became, the more this loss of energy (as the fourth power of energy for relativistic electrons) became a serious matter.[16] After the first direct observation of synchrotron radiation in the late 1940s, particle physicists were interested in the meticulous analysis of the phenomenon of synchrotron radiation because they wanted to counter this obstacle that limited the performance of their accelerators and experiments. Research with synchrotron radiation, in contrast, started to use this radiation as an experimental tool. In other words, the "interest [of particle physics, author's note] in minimizing energy loss ran counter to the prospects of optimizing the emission of synchrotron radiation."[17] In its early decades, research with synchrotron radiation did not possess own, dedicated facilities. But researchers were highly dependent on the good will of particle physicists to share their equipment and instruments, and to offer experimental time at their accelerators. This was all the more problematic because research with synchrotron radiation was at no point to impede the experiments in particle physics (see also Sect. 4.1).[18] Hence, the history of particle physics is not only important with regard to the uneasy relationship between the two groups considering their scientific practices and conduction of experiments. But it is also important with regard to the apparent competition for political commitment and funding and the transformation of the scientific orientation of several laboratories, such as DESY (Deutsches Elektronen Synchrotron, German Electron Synchrotron) or SLAC (Stanford Linear Accelerator Center), from particle physics to photon science in the recent decades (see Sect. 5.1).[19]

The historical development of research with synchrotron radiation, from part-time use of particle physics accelerators to the creation of dedicated synchrotron radiation sources, is commonly classified into three generations. Particle accelerators that were parasitically used for research with synchrotron radiation are called first-generation synchrotron radiation sources. The term *parasitic* means in this context the unplanned use of accelerators, which were originally purpose-built machines for particle physics experiments, for research with synchrotron radiation.[20] Accelerators (storage rings) that became dedicated to research with synchrotron radiation from the late 1970s and early 1980s onwards are called second-generation synchrotron radiation sources. A third generation of synchrotron radiation sources emerged in the late 1980s and went operational in the early 1990s. They had an optimised design and used insertion devices that improved the overall performance of the accelerator and the emitted radiation.[21] Free-electron lasers are often called fourth-generation light sources. This name is, however, misleading. Free-electron lasers are rather complementary to already existing synchrotron radiation sources. But regarding their laser-like characteristics and the implementation of novel technologies, such as linac technologies, they are not quite comparable to existing circular-shaped synchrotron radiation sources.[22]

Early efforts in research with synchrotron radiation took place in the mid 1960s at several places in Europe and the United States, which were originally purpose-built for research with particle physics. These included the Italian Frascati Laboratory, the synchrotron ELSA (Elektronen-Stretcher Anlage) at Bonn University and the ring accelerator DESY, both located in Germany, the Glasgow linear accelerator in the United Kingdom and the 180 MeV electron synchrotron at the National Bureau of Standards in the United States.[23]

The development of storage rings in the late 1960s and 1970s was an important milestone. Storage rings are a specific kind of circular-shaped accelerators: particles can be stored while being kept circulating for several hours providing a more reliable and intense particle beam.[24] Their general design consists of evacuated pipes that are circularly arranged. Electrons are first emitted by a source. They are then accelerated in a linac and further (if needed) in a booster synchrotron. The accelerated electrons are transferred to the large main ring through which they are further accelerated radially close to the speed of light. The particle beam is redirected and disturbed through a magnetic field to stay on its curved path (see Fig. 3.1).[25]

In the beginning, storage rings, too, were only parasitically used for research with synchrotron radiation. But operation dominantly remained in the hands of particle physics in the 1960s and 1970s. The creation of a research programme entirely dedicated to research with synchrotron radiation at the Tantalus storage ring of the University of Wisconsin in the United States in 1968 was a veritable exception in this regard.[26] Other

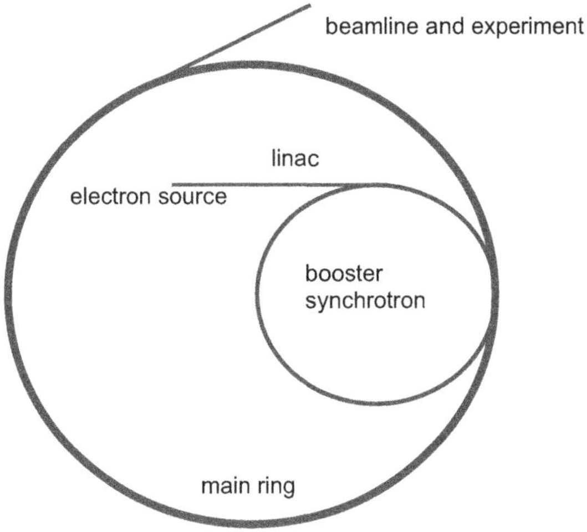

Fig. 3.1 Basic layout of a storage ring. (Note: Figure crafted by the author)

examples of partly used storage rings at that time include DORIS (Doppel-Ring-Speicher, Double Ring Storage) at DESY in Germany or SPEAR (Stanford Positron Electron Asymmetric Rings) at SLAC in the United States. CHESS (Cornell High-Energy Synchrotron Source) at the Cornell University in the United States was closely linked to the particle physics programme at the electron storage ring CESR (Cornell Electron Storage Ring). The ACO (Anneau de Collisions d'Orsay) storage ring in France was partly used as a synchrotron radiation source from the early 1970s onwards when LURE (Laboratoire pour l'Utilisation du Rayonnement Électromagnétique), a laboratory dedicated to research with synchrotron radiation, was created.[27]

The situation changed with what several scholars characterised as the escalation of particle physics research to megascience, with ever-larger experiments paralleled by the increasing isolation and disconnection of particle physics from other scientific fields and experimental research.[28] With regard to the US-American context, the emergence of megascience projects in particle physics coincided with a slowdown of spending increases (albeit not a decline) due to budget cuts in the 1960s and 1970s. At around the same time, an advisory panel of the US-American Atomic Energy Commission had recommended to aim for ever-higher energies in particle physics research, and thus for ever-larger and costlier accelerators. This led to the monopolisation of the national particle physics budget within "one flagship machine at a time."[29] In Europe, the situation was quite similar at this time. The SPS (Super Proton Synchrotron) at CERN (European Organization for Nuclear Research) had gone in operation. The CERN II upgrade in the 1970s led to the cancellation of several national programmes and the channelling of large parts of the national particle physics budgets into CERN. However, these developments towards megascience in Europe and the United States can also be interpreted as the beginning of a process towards dedicated synchrotron radiation sources. The monopolisation of budget and experiments in particle physics research into a few places in Europe and the United States left some (national) accelerators deserted so that research with synchrotron radiation could took over.[30] This was, for instance, the case for efforts at Brookhaven, Cornell University and the University of Wisconsin in the United States, as well as Daresbury in the United Kingdom, Lund in Sweden and West Berlin in West Germany (see below).

The British synchrotron radiation source SRS (Synchrotron Radiation Source) at Daresbury, operating in the UV/soft X-ray spectrum, opened in 1981.[31] In West Germany, BESSY (Berliner Elektronenspeicherring-Gesellschaft für Synchrotronstrahlung mbH) opened in 1981, operating a synchrotron radiation source in the UV/soft X-ray spectrum. HASYLAB (Hamburger Synchrotronstrahlungslabor), as a distinct laboratory for research with synchrotron radiation at DESY, was inaugurated in 1980. Since 1975, LUSY (Lund University Synchrotron) in Lund, Sweden was used for experiments in solid-state physics. The related synchrotron radiation source MAX I (Microtron Accelerator for X-rays) then opened in Lund in 1986. In the United States, the Tantalus storage ring was replaced by the Aladdin source in 1981, fully dedicated to synchrotron radiation and operating in the UV/soft X-ray spectrum. The Brookhaven National Laboratory operated the NSLS (National Synchrotron Light Source) from 1982 onwards. Originally planned and designed in the UV/soft X-ray spectrum, it later expanded operation also to the hard X-ray spectrum.[32] The French national LURE laboratory started operation of SuperACO in 1987. As an upgrade of the already existing ACO storage ring, it now became entirely dedicated to synchrotron radiation.[33] These dedicated efforts are called second-generation synchrotron radiation research. However, most of research with synchrotron radiation remained parasitic until the 1990s.

Insertion devices that are magnetic devices to enhance the quality of the electron beam and hence the emitted radiation were another crucial breakthrough for research with synchrotron radiation. Wigglers and undulators are two kinds of insertion devices that were introduced around the 1970s and 1980s. They improved the performance of bending magnets that were used in early synchrotrons in the 1950s and 1960s. Both consist of an array of dipole magnets with alternating polarity. They bend the electron beam back and forth. The beam thus wiggles or undulates in the horizontal direction. The difference between wigglers and undulators is only of quantitative nature, regarding their field strength that leads, however, to remarkable differences in the quality of the emitted radiation.[34]

Three major third-generation synchrotron radiation sources that make use of insertion devices and that operate in the hard X-ray regime went into operation in the 1990s: the ESRF (European Synchrotron Radiation Facility), the APS (Advanced Photon Source) at Argonne National Laboratory in the United States and the SPring-8 (Super Photon Ring-8 GeV) facility in Japan. All three facilities are based on accelerators with large circumferences that required large investments to be constructed. These three facilities are popularly called the *big three* to highlight their competitive standing compared to other synchrotron radiation sources, namely the similarities in their outstanding performance parameters and their large capacities to serve a great number of users every year. The 1990s also saw the establishment of several smaller, third-generation facilities operating in the UV and soft X-ray range. Most of them were conceptualised and operated as national facilities. Among them was the Elettra synchrotron in Trieste, Italy and the ALS (Advanced Light Source) at Lawrence Berkeley National Laboratory in the United States. Both opened to users in 1993. DORIS III at DESY became a dedicated synchrotron radiation source in 1993. DELTA (Dortmund Electron Accelerator) at the University of Dortmund, Germany opened in 1995. The electron storage ring BESSY II opened in 1998. MAX II opened in 1997. Both enhance and extend the performances of their predecessors: MAX I and BESSY I (see above).[35]

There are some exceptions that do not fit the principle of the three generations: First, intermediary storage rings that started operation by the early 2000s and that are a specific type of third-generation sources improving and refining several parameters of conventional third-generation sources.[36] They have successively opened since the early 2000s in different places in Europe and the United States, such as SLS (Swiss Light Source) at the PSI (Paul Scherrer Institute) in Villingen, Switzerland, ANKA (Angströmquelle Karlsruhe) at the KIT (Karlsruhe Institute of Technology), Germany, SOLEIL (Source Optimisée de Lumière d'Énergie Intermédiaire du LURE) in France, Diamond in the United Kingdom and the Alba light source near Barcelona in Spain. The upgrade of the Swedish facility MAX to MAX IV, as well as the upgrade of SLAC's SPEAR into SPEAR 3, are also among these intermediary light sources.[37] Second, PETRA III (Positron-Elektron Tandem Ring Anlage) at DESY can also be characterised as an exception with regard to its extremely low emittance and high brilliance.[38]

From the 1980s onwards, linear accelerators with laser-like characteristics[39] were expected to perform as next-generation light sources.[40] Higher brightness and coherence, as well as lower emittance and shorter pulse lengths, were scientifically needed to advance experimental investigations into the structure and dynamics of matter. However, these parameters depended on the performance and quality of the electron beam, and hence the design of the source, accelerator and magnetic devices.

Generally speaking, the intensity of the emitted synchrotron radiation is a fundamental parameter for all experiments at synchrotron radiation sources and free-electron lasers. Intensity is synonymous to brilliance or brightness that is the number of photons in a specific volume of the beam. It is thus a measurement of the concentration of the radiation. The brilliance of synchrotron radiation depends on the beam's transverse size and divergence that is termed emittance. Decreasing emittance results, simply speaking, in increasing brightness: The narrower the transverse size and divergence of the beam, the more photons are compressed into a specific volume of the beam.[41]

At the beginning of the 1990s, the circular design of operating synchrotron radiation facilities such as the *big three* (ESRF, APS and SPring-8) put a limit to the further enhancement of these performance parameters.[42] This limit could be challenged by technological innovations (e.g. magnet design or emittance reduction) over the course of the last decades. Intermediary third-generation light sources could then be upgraded far beyond what was initially thought feasible back in the 1990s. At this time and in this context, free-electron lasers based on a linear[43] accelerator design followed by long arrays of undulators were nonetheless expected to operate as next-generation light sources and to provide both very short pulses and brilliant and coherent light in the X-ray wavelength. This was then successfully demonstrated by the operation of FLASH (Free Electron

Laser in Hamburg) at DESY and the LCLS (Linac Coherent Light Source) at SLAC.[44]

The SASE (Self-Amplified Spontaneous Emission) principle is of fundamental importance for X-ray free-electron laser. For a long time, it has been unclear how a laser could achieve coherent light in the VUV and hard and soft X-ray regions.[45] Coherent light in these wavelength regions was, however, of crucial importance for the performance of experiments. While conventional gas or solid-state lasers could operate in the infrared, visible and ultraviolet range using optic cavities (e.g. mirrors) to build up the radiation, this was not possible for the shorter wavelengths in the vacuum ultraviolet or soft and hard X-ray regions. Because the radiation would have damaged these mirrors, here SASE mechanism had to play an important role: when the undulators cause the electron beam to emit radiation, the radiation travels faster than the electrons. The electron beam then interacts with its own electromagnetic field while further moving through the long undulator. This process thus creates a constant phase relation and constructive interference between the electrons and produces, importantly, coherent light. It was introduced by the physicists Evgeny L. Saldin and Anatoli Kondratenko,[46] as well as Rodolfo Bonifacio, Claudio Pellegrini and Lorenzo Narducci[47] in the 1980s and paved the way to the construction of free-electron lasers in the regions of X-ray and vacuum ultraviolet.

Early theoretical work on free-electron lasers was done throughout the 1960s and 1970s at the same time when first experiments with synchrotron radiation were set up. However, for a long time, these lasers could only operate in a very narrow slice of the spectral range (for the reasons mentioned above), whereas synchrotron radiation was available and tuneable in a very broad and continuous slice of the electromagnetic spectrum (see above). This is also one reason why synchrotron radiation sources remained the preferred experimental tools, although early versions of free-electron lasers already surpassed the performance of circular-shaped synchrotron radiation sources in terms of brilliance, and offered coherent light.[48]

In free-electron lasers, electrons travel like particles at relativistic velocity. Electrons are thus set free. They are not bound to matter, as this is the case for conventional lasers.[49] The basic design of free-electron lasers consists of a source that is followed by a linac and a long array of undulators (see Fig. 3.2). While in many textbooks and handbooks, the magnetic structures of free-electron lasers are generally called undulators, some

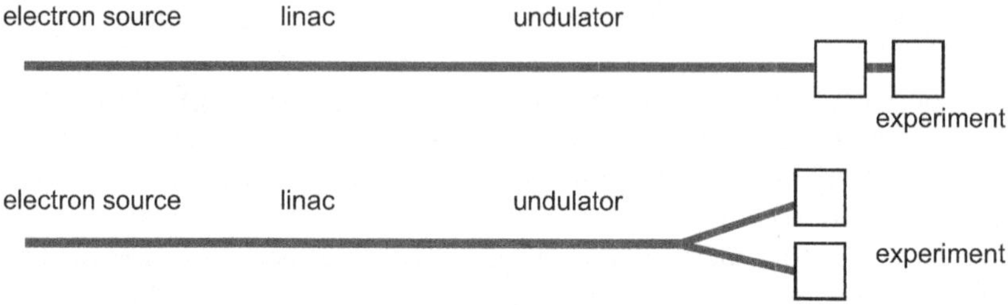

Fig. 3.2 Basic layouts of free-electron lasers. (Note: Figure crafted by the author)

scientists argue that a distinction is "arbitrary" and that "'wiggler' is generally used to describe the periodic magnets in free-electron lasers, while 'undulator' is used for incoherent synchrotron light sources."[50] As free-electron lasers have a linear design, the beam is lost after one pass because there is no way of a second passing, as is the case in the closed loops of circular-shaped synchrotrons.[51]

The first optical free-electron laser, tuneable over a broad wavelength spectrum, was successfully put into operation in 1982, and was based on the ACO storage ring at the University Paris Sud, France.[52] The VEPP-3 (Russian acronym for "Colliding Electron Beams 3") storage ring in Novosibirsk in the Soviet Union reported pioneering operation of its free-electron laser in the visible and UV wavelength ranges in 1988.[53]

In Japan, a free-electron laser in the VUV spectrum went into operation in 2008 at the Spring-8 facility. The construction of SACLA (SPring-8 Angstrom Compact Free Electron Laser) at the same facility was completed in 2011. Since 2012 it provides a stable beam for user operation in the Ångström wavelength region.[54] China built the SDUV-FEL (Shanghai Deep-Ultraviolet Free Electron Laser), which began operation in 2010 at SINAP (Shanghai Institute of Applied Physics). SINAP also hosts the SXFEL (Shanghai Soft X-rays Free Electron Laser), which is currently turned into a user facility.[55]

In the United States, the LCLS at SLAC operates in the hard X-ray spectrum. In 2009, LCLS was the first tuneable free-electron laser that achieved femtosecond pulses in the Ångström wavelength. An upgrade of the LCLS (into LCLS-II) is expected to provide ultrafast X-rays in the near future.[56] The Thomas Jefferson National Accelerator Facility operates a free-electron laser in the IR spectrum. The Berkeley National Laboratory has established a new design for a free-electron laser in the soft X-ray region, making the parallel operation of several user groups possible. The Cornell University currently develops an alternative type of light source, an ERL (Energy Recovery Linac) for the hard X-ray region.[57]

In Germany, FLASH at DESY opened to external users in 2005. It was also at FLASH that the SASE principle at short wavelengths was experimentally observed and confirmed for the first time (see above and Sect. 5.3).[58] A proposal of the BESSY II facility in Berlin, Germany to build a free-electron laser in the soft X-ray spectrum (BESSY-FEL) was abandoned in 2008. Funders had decided that this proposal should upgrade FLASH at DESY, while BESSY II should begin the preparation of an alternative kind of free-electron laser, which is an ERL. In Italy, FERMI (Free Electron laser Radiation for Multidisciplinary Investigations) at Elettra started user operation in the extreme UV and soft X-ray range.[59]

Not only particle physics (see above) but also research with neutrons and ions constitute important and additional scientific contexts to research with synchrotron radiation because they can be regarded as complementary experimental resources. Research with synchrotron radiation (most notably in the X-ray regime), neutrons and ions are (among several others) powerful tools to study the structure and dynamics of matter. Depending on the intention and purpose of the experiment, they offer complementary insights and results. The choice of X-ray, neutrons and/or ions thus depends on the composition of the material under investigation, as well as the kind of structural information needed. By using different techniques, all three resources can reveal fundamental information on the properties and characteristics of a broad variety of samples such as organic or inorganic matter, drugs, plasma or liquids.[60] To illustrate this point, in 2017, the journal *Quantum Beam Science* (QuBS) was launched with the intention to specifically address the complementarity between research with neutrons, ions and synchrotron radiation, as well as several other experimental resources. Editor Klaus-Dieter Liss pointed out that "[q]uantum beams encompass all kinds of short-wavelength radiation for the study of condensed matter materials in the broadest sense,"[61] including the abovementioned synchrotron radiation, neutrons and ion radiation but also positrons or muons. "While most of those quantum beams can be generated on a small laboratory scale," Liss continues, "state-of-the-art sources are assembled in large-scale multi-user facilities, such as spallation sources, third-generation synchrotrons and nuclear reactors."[62] It thus needs to be retained that the history of research with neutrons and ions provides complementary, and sometimes competing, backgrounds to study and understand the historical development of research with synchrotron radiation. Hence, the creation of projects such as the neutron reactor of the ILL (Institut Laue-Langevin) in Grenoble, France, which is co-located with the ESRF, the antiproton and ion accelerator of the FAIR (Facility for Antiproton and Ion Research) facility in Darmstadt, Germany, and the neutron spallation source ESS (European Spallation Source), which is under construction in Lund, Sweden, and also somewhat smaller projects such as the British ISIS project and the German SNQ (Spallations-Neutronenquelle) project, which was planned in the 1970s and 1980s but never realised, needs to be considered equally (see Sects. 4.2 and 5.4).

Notes

1. See, for example, H. Pollock, "The Discovery of Synchrotron Radiation." *American Journal of Physics* 51, no. 3 (1983); H. Winick and S. Doniach, "An Overview of Synchrotron Radiation Research." In *Synchrotron Radiation Research*, eds. H. Winick and S. Doniach (Boston, MA: Springer, 1980), 4.
2. See, for example, A.-M. Liénard, "Champ Électrique et Magnétique Produit par une Charge Électrique Concentrée en un Point et Animée d'un Mouvement Quelconque" *L'Éclairage Électrique* 16, no. 27 (1898).
3. See, for example, C. Roychoudhuri, A. F. Kracklauer, and K. Creath, *The Nature of Light: What is a Photon?* (Boca Raton: CRC Press, 2008).
4. See, for example, K.-D. Liss and K. Chen, "Frontiers of Synchrotron Research in Materials Science." *MRS Bulletin* 41, no. 6 (2016).
5. See, for example, R. Blaustein, "Biology and Light Sources: Synchrotrons Allow Researchers a Deep Look into Life." *BioScience* 67, no. 3 (2017).
6. See, for example, W. Thomlinson, P. Suortti, and D. Chapman, "Recent Advances in Synchrotron Radiation Medical Research." *Nuclear Instruments and Methods in Physics Research, Section A* 543, No. 1 (2005).
7. See, for example, European XFEL, *Enlightening Science* (Schenefeld: European XFEL GmbH, 2017).
8. See, for example, A. Adriaens, "Non-Destructive Analysis and Testing of Museum Objects: An Overview of 5 Years of Research." *Spectrochimica Acta Part B* 60, no. 12 (2005).
9. See, for example, M. Fox, *Quantum Optics: An Introduction* (Oxford, New York: Oxford University Press, 2006), 13–16.
10. See, for example, H.-D. Nuhn, "From Storage Rings to Free Electron Lasers for Hard X-Rays." *Journal of Physics: Condensed Matter* 16, no. 33 (2004), 16.
11. See C. Kunz, "Introduction: Properties of Synchrotron Radiation." In *Synchrotron Radiation: Techniques and Applications*, ed. C. Kunz (Berlin, Heidelberg: Springer Verlag, 1979).

12. The electromagnetic spectrum maps and classifies electromagnetic radiation according to its wavelength. It ranges from gamma rays with a wavelength of 1 pm to extremely low frequency radio waves with a wavelength up to 100,000 km.
13. See O. Hallonsten and T. Heinze, "Formation and Expansion of a New Organizational Field in Experimental Science." *Science and Public Policy* 42, no. 6 (2015), 845.
14. See O. Hallonsten, *Big Science Transformed: Science, Politics and Organization in Europe and the United States* (Cham: Palgrave Macmillan, 2016), 241–242, 249.
15. See O. Hallonsten, *Big Science Transformed*, 241–242, 249.
16. See Pollock, "The Discovery", 278.
17. Hallonsten and Heinze, "Formation and Expansion", 844.
18. See O. Hallonsten, "The Parasites." *Historical Studies in the Natural Sciences* 45, no. 2 (2015).
19. See, for example, O. Hallonsten and T. Heinze, "From Particle Physics to Photon Science: Multidimensional and Multi-Level Renewal at DESY and SLAC." *Science and Public Policy* 40, no. 5 (2013).
20. See Hallonsten, "The Parasites".
21. See, for example, M. Altarelli, "Physics with Third Generation Synchrotron Sources." *Physica Scripta* T55 (1994).
22. See H. Winick, *Fourth Generation Light Sources*, paper presented at the17th IEEE Particle Accelerator Conference (PAC 97), Vancouver, Canada (May 12–16, 1997).
23. See, for example, C. Kunz, M. Skibowski and B. Sonntag, "How It All Started at DESY in 1964." *Synchrotron Radiation News* 28, no. 4 (2015); P. Dhez, "Synchrotron Radiation in France: The Early Years." *Synchrotron Radiation News* 28, no. 4 (2015); T. Heinze, O. Hallonsten and S. Heinecke, "From Periphery to Center: Synchrotron Radiation at DESY, Part I: 1962–1977." *Historical Studies in the Natural Sciences* 45, no. 3 (2015); C. Kunz, *Synchrotronstrahlung bei DESY: Anfänge* (private print, 2012).
24. V. Vylet and J. Liu, *Synchrotron Facilities and Free Electron Lasers*, SLAC-PUB-13049 (Stanford: SLAC, 2007), 6.
25. See H. Winick, "Properties of Synchrotron Radiation." In *Synchrotron Radiation Research*, eds. H. Winick and S. Doniach (Boston, MA: Springer, 1980), 20; Vylet and Liu, *Synchrotron Facilities and Free Electron Lasers*, 6.
26. See D. Lynch et al., "Tantalus, the First Dedicated Synchrotron Radiation Source." *Synchrotron Radiation News* 28, no. 4 (2015).
27. See, for example, Winick and Doniach, "An Overview", 5; Hallonsten and Heinze, "Formation and Expansion", 4; A. Michalowicz, H. Ostrowiecki,

and I. Martelly, "Bref Historique sur les Instruments." *Histoire de la Recherche Contemporaine* 3, no. 1 (2014); Hallonsten and Heinze, "From Particle Physics to Photon Science"; Heinze, Hallonsten and Heinecke, "From Periphery to Center I".

28. See, for example, L. Hoddeson, A. Kolb and C. Westfall, *Fermilab: Physics, the Frontier, and Megascience* (Chicago: University of Chicago Press, 2008), 2–3; Hallonsten, *Big Science Transformed*, 63 f.
29. See Hallonsten, *Big Science Transformed*, 63.
30. See Hallonsten, *Big Science Transformed*, 80; Cramer, "Lightening Europe: Establishing the European Synchrotron Radiation Facility (ESRF)." *History and Technology* 33, no. 4 (2017), 399.
31. See I. Munro, "Fifty Years of Synchrotron Radiation Research in the UK." *Philosophical Transactions of the Royal Society* 377, no. 2147 (2019).
32. See, for example, O. Hallonsten, "Growing Big Science in a Small Country: MAX-Lab and the Swedish Research Policy System." *Historical Studies in the Natural Sciences* 41, no. 2 (2011); Hallonsten and Heinze, "Formation and Expansion", 3; Cramer, "Lightening Europe", 406.
33. See M. Belakhovsky, "Histoire et Développement des Sources de Rayonnement Synchrotron." *Reflets de la Physique* 34–35 (2013).
34. See, for example, Nuhn, "From Storage Rings", 15; W. Eberhardt, "Synchrotron Radiation: A Continuing Revolution in X-Ray Science – Diffraction Limited Storage Rings and Beyond." *Journal of Electron Spectroscopy and Related Phenomena* 200 (2015).
35. See, for example, E. Lohrmann and P. Söding, *Von schnellen Teilchen und hellem Licht: 50 Jahre Deutsches Elektronen-Synchrotron DESY* (Weinheim: Wiley, 2009); T. Heinze, O. Hallonsten and S. Heinecke, "From Periphery to Center: Synchrotron Radiation at DESY, Part II: 1977–1993." *Historical Studies in the Natural Sciences* 45, no. 4 (2015); Hallonsten and Heinze, "Formation and Expansion", 845 and Table 1.
36. See Hallonsten and Heinze, "Formation and Expansion", 846.
37. See Hallonsten and Heinze, "Formation and Expansion", Table 1.
38. See K. Balewski et al., eds., *PETRA III: A Low Emittance Synchrotron Radiation Source: Technical Design Report*, Executive Summary (Hamburg: DESY, 2004).
39. There are mainly two kinds of linac-based light sources with the characteristics of a laser: FEL (free-electron laser) and ERL (energy recovery linac). Only the FEL will be discussed in detail here.
40. See, for example, H. Winick, *Synchrotron Radiation Sources: Present Capabilities and Future Directions*, SLAC-PUB-777 1 (Stanford, 1998), 1; Nuhn, "From Storage Rings", 16.

41. See, for example, Nuhn, "From Storage Rings".
42. See G. Materlik and T. Tschentscher, eds., *TESLA Technical Design Report: Part V: The XFEL* (Hamburg: DESY, 2001), 13.
43. There also exist free-electron lasers based on circular storage rings.
44. See J. Feldhaus, J. Arthur and J. Hastings, "X-Ray Free-Electron Lasers." *Journal of Physics B* 38, no. 9 (2005), 799.
45. See C. Pellegrini, "The History of X-Ray Free-Electron Lasers." *The European Physical Journal H* 37, no. 5 (2012).
46. See A. Kondratenko and E. Saldin, "Generation of Coherent Radiation by a Relativistic Electron Beam in an Ondulator." *Particle Accelerators* 10 (1980).
47. See R. Bonifacio, C. Pellegrini, and L. Narducci, "Collective Instabilities and High-Gain Regime in a Free Electron Laser." *Optics Communications* 50, no. 6 (1984).
48. See, for example, C. Pellegrini, "X-Ray Free-Electron Lasers: From Dream to Reality." *Physica Scripta* T169 (2016).
49. See Z. Huang and P. Schmüser, "Free-Electron Lasers." In *Handbook of Accelerator Physics and Engineering*, eds. A. Chao et al. (Singapore: World Scientific, 2013), 229.
50. See, for example, H. Freund and T. Antonsen, *Principles of Free-Electron Lasers* (Dordrecht: Springer Netherlands, 1992), 2.
51. See Vylet and Liu, *Synchrotron Facilities*, 20.
52. See M. Billardon et al., "First Operation of a Storage-Ring Free-Electron Laser." *Physical Review Letters* 51, no. 18 (1983).
53. See I. Drobyazko et al., "Lasing in Visible and Ultraviolet Regions in an Optical Klystron Installed on the VEPP-3 Storage Ring." *Nuclear Instruments and Methods in Physics Research Section A* 282, 2–3 (1989).
54. See M. Yabashi et al., "Status of the SACLA Facility." *Applied Sciences* 7, no. 6 (2017).
55. See Z. Zhentang et al., "Status of the SXFEL Facility." *Applied Sciences* 7, no. 6 (2017).
56. See R. Schoenlein et al., "The Linac Coherent Light Source: Recent Developments and Future Plans." *Applied Sciences* 7, no. 8 (2017).
57. See Komitee für Forschung mit Synchrotronstrahlung. *Forschung mit Synchrotronstrahlung in Deutschland: Status und Perspektiven* (2009), 9; Z. Zhao and D. Wang, "Seeded FEL Experiments at the SDUV-FEL Test Facility." *IEEE Transactions on Nuclear Science* 63, no. 2 (2016).
58. See J. Andruszkow et al., "First Observation of Self-Amplified Spontaneous Emission in a Free-Electron Laser at 109 nm Wavelength." *Physical Review Letters* 85, no. 18 (2000).

59. L. Giannessi and C. Masciovecchio, "FERMI: Present and Future Challenges." *Applied Sciences* 7, no. 6 (2017).
60. See, for example, K.-D. Liss, "Quantum Beam Science – Applications to Probe or Influence Matter and Materials." *Quantum Beam Science* 1, no. 1 (2017); K.-D. Liss and K. Chen, "Frontiers of Synchrotron Research".
61. K.-D. Liss, "Facilities in Quantum Beam Science." *Quantum Beam Science* 2, no. 6 (2018), 1.
62. Liss, "Facilities in Quantum Beam Science", 1.

Bibliography

Adriaens, A. "Non-Destructive Analysis and Testing of Museum Objects: An Overview of 5 Years of Research." *Spectrochimica Acta Part B* 60, no. 12 (2005): 1503–1516.

Altarelli, M. "Physics with Third Generation Synchrotron Sources." *Physica Scripta* T55 (1994): 9–13.

Andruszkow, J., Aune, B., Ayvazyan, V., Baboi, N., Bakker, R., Balakin, V., Barni, D., Bazhan, A., Bernard, M., Bosotti, A., et al. "First Observation of Self-Amplified Spontaneous Emission in a Free-Electron Laser at 109 nm Wavelength." *Physical Review Letters* 85, no. 18 (2000): 3825–3829.

Balewski, K., Brefeld, W., Decking, W., Franz, H., Röhlsberger, R. and Weckert, E., eds. *PETRA III: A Low Emittance Synchrotron Radiation Source: Technical Design Report* (Executive Summary). Hamburg: DESY, 2004.

Belakhovsky, M. "Histoire et Développement des Sources de Rayonnement Synchrotron." *Reflets de la Physique*, 34–35 (2013): 10–11.

Billardon, M., Elleaume, P., Ortega, J., Bazin, C., Bergher, M., Velghe, M., Petroff, Y., Deacon, D., Robinson, K., Madey, J. "First Operation of a Storage-Ring Free-Electron Laser." *Physical Review Letters* 51, no. 18 (1983): 1652–1655.

Blaustein, R. "Biology and Light Sources: Synchrotrons Allow Researchers a Deep Look into Life." *BioScience* 67, no. 3 (2017): 201–207.

Bonifacio, R., Pellegrini, C., and Narducci, L. "Collective Instabilities and High-Gain Regime in a Free Electron Laser." *Optics Communications* 50, no. 6 (1984): 373–378.

Cramer, K. C. "Lightening Europe: Establishing the European Synchrotron Radiation Facility (ESRF)." *History and Technology* 33, no. 4 (2017): 396–427.

Dhez, P. "Synchrotron Radiation in France: The Early Years." *Synchrotron Radiation News* 28, no. 4 (2015): 42–43.

Drobyazko, I., Kulipanov, G., Litvinenko, V., Pinayev, I., Popik, V., Silvestrov, I., Skrinsky, A., Sokolov, A., and Vinokurov, N. "Lasing in Visible and Ultraviolet Regions in an Optical Klystron Installed on the VEPP-3 Storage Ring." *Nuclear Instruments and Methods in Physics Research Section A* 282, 2–3 (1989): 424–430.

Eberhardt, W. "Synchrotron Radiation: A Continuing Revolution in X-Ray Science – Diffraction Limited Storage Rings and Beyond." *Journal of Electron Spectroscopy and Related Phenomena* 200 (2015): 31–39.

European XFEL, *Enlightening Science* (Schenefeld: European XFEL GmbH, 2017).

Feldhaus, J., Arthur, J., and Hastings, J. "X-Ray Free-Electron Lasers." *Journal of Physics B* 38, no. 9 (2005): 799–819.

Fox, M. *Quantum Optics: An Introduction.* Oxford, New York: Oxford University Press, 2006.

Freund, H. and Antonsen, T. *Principles of Free-Electron Lasers.* Dordrecht: Springer Netherlands, 1992.

Giannessi, L. and Masciovecchio, C. "FERMI: Present and Future Challenges." *Applied Sciences* 7, no. 6 (2017): 640.

Hallonsten, O. "Growing Big Science in a Small Country: MAX-Lab and the Swedish Research Policy System." *Historical Studies in the Natural Sciences* 41, no. 2 (2011): 179–215

Hallonsten, O. "The Parasites." *Historical Studies in the Natural Sciences* 45, no. 2 (2015): 217–272.

Hallonsten, O. *Big Science Transformed: Science, Politics and Organization in Europe and the United States.* Cham: Palgrave Macmillan, 2016.

Hallonsten, O. and Heinze, T. "Formation and Expansion of a New Organizational Field in Experimental Science." *Science and Public Policy* 42, no. 6 (2015): 841–854.

Hallonsten, O. and Heinze, T. "From Particle Physics to Photon Science: Multidimensional and Multi-Level Renewal at DESY and SLAC." *Science and Public Policy* 40, no. 5 (2013): 591–603.

Heinze, T., Hallonsten, O., and Heinecke, S. "From Periphery to Center: Synchrotron Radiation at DESY, Part I: 1962–1977." *Historical Studies in the Natural Sciences* 45, no. 3 (2015a): 447–492.

Heinze, T., Hallonsten, O., and Heinecke, S. "From Periphery to Center.: Synchrotron Radiation at DESY, Part II: 1977–1993." *Historical Studies in the Natural Sciences* 45, no. 4 (2015b): 513–548.

Hoddeson, L., Kolb, A., and Westfall, C. *Fermilab: Physics, the Frontier, and Megascience.* Chicago: University of Chicago Press, 2008.

Huang, Z. and Schmüser, P. "Free-Electron Lasers." In *Handbook of Accelerator Physics and Engineering*, edited by A. W. Chao, K. H. Mess, T. Maury, and F. Zimmermann, 229–233. Singapore: World Scientific, 2013.

Komitee für Forschung mit Synchrotronstrahlung. *Forschung mit Synchrotronstrahlung in Deutschland: Status und Perspektiven.* 2009. Online available: https://www.sni-portal.de/de/Dateien/forschung-mit-synchrotronstrahlung-in-deutschland-status-und-perspektiven-2/at_download/file, last accessed 1 March 2020.

Kondratenko, A. and Saldin, E. "Generation of Coherent Radiation by a Relativistic Electron Beam in an Ondulator." *Particle Accelerators* 10 (1980): 207–216.
Kunz, C. "Introduction: Properties of Synchrotron Radiation." In *Synchrotron Radiation: Techniques and Applications*, edited by C. Kunz, 1–23. Berlin, Heidelberg: Springer Verlag 1979.
Kunz, C. *Synchrotronstrahlung bei DESY: Anfänge.* Private print, 2012.
Kunz, C., Skibowski, M., and Sonntag, B. "How It All Started at DESY in 1964." *Synchrotron Radiation News* 28, no. 4 (2015): 16–19.
Liénard, A.-M. "Champ Électrique et Magnétique Produit par une Charge Électrique Concentrée en un Point et Animée d'un Mouvement Quelconque" *L'Éclairage Électrique* 16, no. 27 (1898): 5–14.
Liss, K.-D. "Facilities in Quantum Beam Science." *Quantum Beam Science* 2, *no.* 6 (2018): 1–4.
Liss, K.-D. "Quantum Beam Science – Applications to Probe or Influence Matter and Materials." *Quantum Beam Science* 1, no. 1 (2017): 1–5.
Liss, K.-D. and Chen, K. "Frontiers of Synchrotron Research in Materials Science." *MRS Bulletin* 41, no. 6 (2016): 435–441.
Lohrmann, E. and Söding, P. *Von schnellen Teilchen und hellem Licht: 50 Jahre Deutsches Elektronen-Synchrotron DESY.* Weinheim: Wiley, 2009.
Lynch, D., Plummer, W., Himpsel, F., Chiang, T., Margaritondo, G., and Lapeyre, G. "Tantalus, the First Dedicated Synchrotron Radiation Source." *Synchrotron Radiation News* 28, no. 4 (2015): 20–23.
Materlik, G. and Tschentscher, T., eds. *TESLA Technical Design Report: Part V: The XFEL.* Hamburg: DESY, 2001.
Michalowicz, A., Ostrowiecki, H. and Martelly, I. "Bref Historique sur les Instruments." *Histoire de la Recherche Contemporaine* 3, no. 1 (2014): 33–36.
Munro, I. "Fifty Years of Synchrotron Radiation Research in the UK." *Philosophical Transactions of the Royal Society* 377, no. 2147 (2019). Online available: https://royalsocietypublishing.org/doi/pdf/10.1098/rsta.2018.0230, last accessed 20 March 2020.
Pellegrini, C. "The History of X-Ray Free-Electron Lasers." *The European Physical Journal* H 37, no. 5 (2012): 659–708.
Pellegrini, C. "X-Ray Free-Electron Lasers: From Dream to Reality." *Physica Scripta* T169 (2016): 1–22.
Pollock, H. "The Discovery of Synchrotron Radiation." *American Journal of Physics* 51, no. 3 (1983): 278–280.
Roychoudhuri, C., Kracklauer, A. F., and Creath K. *The Nature of Light: What is a Photon?* Boca Raton: CRC Press, 2008.
Schoenlein, R., Boutet, S., Minitti, M., and Dunne, A. "The Linac Coherent Light Source: Recent Developments and Future Plans." *Applied Sciences* 7, no. 8 (2017): 850.

Vylet, V. and Liu, J. *Synchrotron Facilities and Free Electron Lasers*, SLAC-PUB-13049. Stanford: SLAC, 2007.

Winick, H. "Properties of Synchrotron Radiation." In *Synchrotron Radiation Research*, edited by H. Winick and S. Doniach, 11–27. Boston, Mass: Springer, 1980.

Winick, H. *Fourth Generation Light Sources*. Paper presented at the 17th IEEE Particle Accelerator Conference (PAC 97); Vancouver, Canada, May 12–16, 1997.

Winick, H. *Synchrotron Radiation Sources: Present Capabilities and Future Directions*, SLAC-PUB-777 1. Stanford: SLAC, 1998.

Winick, H. and Doniach, S. "An Overview of Synchrotron Radiation Research." In *Synchrotron Radiation Research*, edited by H. Winick and S. Doniach, 1–10. Boston, Mass: Springer, 1980.

Yabashi, M., Tanaka, H., Tono, K., and Ishikawa, T. "Status of the SACLA Facility." *Applied Sciences* 7, no. 6 (2017): 604.

Zhentang, Z., Dong, W., Qiang, G., Lixin, Y., Ming, G., Yongbin, L., and Bo, L. "Status of the SXFEL Facility." *Applied Sciences* 7, no. 6 (2017): 607.

CHAPTER 4

Founding the European Synchrotron Radiation Facility (ESRF), 1977–1988

The ESRF was the first collaborative synchrotron radiation facility in Europe. It was established in 1988 as a limited liability company under French domestic law (*société civile*) based on an intergovernmental agreement among eleven European countries: Belgium, Denmark, Finland, France, Germany,[1] Italy, Norway, Spain, Sweden, Switzerland and the United Kingdom. The ESRF is based in Grenoble, France on a peninsula that is called the *Polygone Scientifique* in French. It is co-located with the ILL (Institut Laue-Langevin) and its high flux neutron research reactor. At the time of its establishment, the ESRF was also surrounded by the CNRS (Centre National de la Recherche Scientifique), CEA (Commissariat à l'Énergie Atomique), INSERM (Institut National de la Santé et de la Recherche Médicale), CNET (Centre National d'Études des Télécommunications) and parts of Grenoble University. The EMBL (European Molecular Biology Laboratory), headquartered in Heidelberg, Germany, had established an outstation at the ESRF.[2] In 1994, after completion of the construction phase, the ESRF opened to external users to carry out experiments.

4.1 Origins of the ESRF

The ESF (European Science Foundation) became the institutional setting in which the first steps towards the realisation of the ESRF were taken. The ESF was founded in 1974 following a proposal by Altiero Spinelli,

K. C. Cramer, *A Political History of Big Science*,
Palgrave Studies in the History of Science and Technology,
https://doi.org/10.1007/978-3-030-50049-8_4

then commissioner at the European Commission. Initial plans for the ESF foresaw a great autonomy with own financial resources from the budget of the EEC (European Economic Community) or from special contributions from public and/or private bodies.[3] The national funding agencies of the European member countries such as the Max Planck Society in Germany or the Royal Society in the United Kingdom had, however, been alarmed by this idea. They claimed that fundamental science, together with its funding schemes and collaborative frameworks, should be driven "by the needs of the science rather than by political agendas,"[4] and that the EEC with only nine members at that time, and the strong focus on economic issues, did not provide an adequate frame for this kind of collaboration. The ESF was eventually established through a rather loosely coupled intergovernmental arrangement: It became organised as an association of national research agencies, but did not have any decision power or funding agency.[5] This point is important to bear in mind because decision on a site and initial funding for the ESRF project escalated beyond the competences of the ESF, and called for joint action on the governmental levels among the potential member countries of this facility.

In 1976, the ESF decided to set up a working group on synchrotron radiation based on recommendations from William Garton, professor at the Imperial College in London, that he had sent in 1975 to Sir Brian Flowers, then president of the ESF and rector of the Imperial College.[6] The main aim of this working group was to improve collaboration in research with synchrotron radiation and to advice on further activities. Next to Heinz Maier-Leibnitz, the chair of this working group and former director of the ILL, the group was composed of, among others, Manuel Cardona, founding director of the German Max Planck Institute for Solid State Research, Yves Farge, first director of the French laboratory LURE (Laboratoire pour l'Utilisation du Rayonnement Électromagnétique), William Garton as well as Ian Munro, founding member of the British SRS (Synchrotron Radiation Source).[7]

At around the same time, the US-American Academy of Sciences had published a report that provided an overview of past and present national activities related to research with synchrotron radiation. This report recommended to increase investments into research with synchrotron radiation as well as to strengthen political support. Several members of the ESF working group started to prepare a similar report that was eventually published in 1977.[8] This report, entitled *Synchrotron Radiation: A Perspective View for Europe*, constituted a key document in the early history of the

ESRF.[9] It pointed to the growing interest of the scientific communities in Europe engaged in research with synchrotron radiation and highlighted the competitive developments in Japan, the Soviet Union and the United States.[10] The report also recommended "[a] large effort to build a dedicated hard x-ray storage ring and appropriate advanced instrumentation with a design which goes beyond that of present day projects."[11] When the members of the executive council of the ESF and the general assembly met in November 1977 to discuss this report and to decide on further actions, they did not had time to read it beforehand.[12] However:

> due to the weight Heinz Maier-Leibnitz put behind it, it was agreed that the SR working group should be transformed and extended into an ad-hoc Committee (...) and that it should carry out the feasibility study for a European Synchrotron Radiation Laboratory.[13]

This short anecdote can be interpreted as symptomatic of the way individual personalities shaped the early history of the ESRF. It was not only Heinz Maier-Leibnitz but also the members of the working group on synchrotron radiation and similar committees (e.g. the progress committee, see below) as well as William Garton who seemingly had enough power and expertise to trigger initial steps towards the realisation of the ESRF. The feasibility study, prepared by an ad-hoc committee chaired by Yves Petroff, was then published in 1979.[14] It reaffirmed the recommendations of the 1977 report, namely, to start preparations for a collaborative synchrotron radiation source in Europe because such a facility will "keep Europe in an outstanding position in the field of synchrotron radiation."[15]

Generally speaking, the mid to late 1970s and early 1980s constituted an important turning point for research with synchrotron radiation, when, for instance, the storage rings SPEAR (Stanford Positron Electron Asymmetric Rings) at SLAC (Stanford Linear Accelerator Center) in the United States and DORIS (Doppel-Ring-Speicher, Double-Ring Storage) at DESY (Deutsches Elektronen-Synchrotron, German Electron Synchrotron) in Germany started operation in the mid-1970s (see Chap. 3).[16] Storage rings can be considered as important milestones in the realisation of more reliable synchrotron radiation sources, although most of these sources remained purpose-built for particle physics experiments at that time. On the one hand, part-time use of these storage rings provided scientists with the opportunity to demonstrate scientific excellence in research with synchrotron radiation. By this, they could not only raise the attention of political decision-makers in the ministries and governmental bodies, which would eventually fund new research projects and experiments. But they also circumvented requesting additional financial support and dedicated funding that certainly would have stood in competition with the (limited) budget foreseen for experiments in particle physics at that time. On the other hand, although new experimental opportunities for particle physics were realised, from which research with synchrotron radiation could (partly) benefit, these were few, but very powerful projects. Scientists, as noted in a study led by Manuel Cardona, were concerned that with regard to this limited number of particle physics facilities, a growing demand for experimental time for particle physics research would be answered by a more efficient and thorough use of the beam. This certainly would limit the share of beam time and the experimental opportunities for researchers with synchrotron radiation to a minimum.[17]

This situation, together with the growing community of synchrotron radiation users, put high demands on the existing experimental opportunities for research with synchrotron radiation at, for instance, SLAC and DESY. Several European countries, such as France, Germany, Italy, Sweden and the United Kingdom, started to plan and realise dedicated, national synchrotron radiation sources. These sources had only a very preliminary technological design and performance parameters (regarding, for instance, the wavelength regime in which they operated, or the number of experiments that they could serve simultaneously) compared to what the ESRF later became able to achieve (see Chap. 3). Yet, these early efforts in research with synchrotron radiation were paralleled by increasing political awareness in several European countries for the promises of this emerging scientific field. They also constituted an important point of reference from which more advanced projects, such as the ESRF, could level off.

After the publication of the feasibility study in 1979, the ESF did not have a mandate to further advance the project because decisions on funding and a site for such collaborative Big Science projects still lay in the hands of the national governments, and depended on agreement and compromise at the highest political levels. In 1981, Friedrich Schneider, secretary of the ESF, as well as his successor Hubert Curien, who later became the French minister for research and technology, approached the governments of several European countries inviting them to support this project idea and to work towards its realisation.[18]

Soon after, a progress committee (also called *Group Levaux* after its chairman Paul Levaux, the secretary of the Belgian National Research Fund) was established. It worked as an intergovernmental committee composed of representatives from Belgium, Denmark, France, Finland, Germany, Ireland, Italy, Sweden and the United Kingdom, and met for the first time in January 1982.[19] The committee was later joint by delegations from Austria, the Netherlands, Switzerland and Yugoslavia. Although it was formally independent from any institution of the EEC, such as the Council of Ministers, there nevertheless was a close exchange of information between the committee and EEC-related instances. The European Commission had, for instance, sent a permanent observer to the meetings of the progress committee.[20] On some rare occasions, the ESRF project was also discussed during meetings of the Council of Ministers when the research ministers from the European member countries met. The Council of Ministers, however, did only provide loose suggestions and did not make any binding decisions. Although the progress committee became an important forum to discuss any kind of project-related questions, such as site proposals, funding as well as organisational matters, it could not take a definite decision regarding these matters.[21]

In December 1982, the progress committee published a report that put a spotlight on the significance of the ESRF in the context of other national efforts on research with synchrotron radiation in Europe and in the United States. With regard to the potentially competitive situation between national research centres in Europe and the United States, it was argued that "the purpose of competing centres is not primarily to outdo the other," but "advancements in science are very often the result of a dialogue between equal partners."[22] Europe, it was concluded, thus needs to develop its own cutting-edge research facilities for research with synchrotron radiation that enable a fruitful international but also interdisciplinary exchange between scientific communities.[23]

The progress committee also created an expert-based study group, the ESRP (European Synchrotron Radiation Project, also called *Group Buras* after its chairman, the physicist Bronislaw Buras) that worked at CERN, close to Geneva, Switzerland. One major task of the ESRP group was to prepare a site-independent proposal for the ESRF and to update the project's technological and scientific design, for which basic parameters had been formulated in the 1979 feasibility study. This work resulted in the so-called *Green Book*, published in 1984.[24] Another task was to explore the potential of synchrotron radiation for industrial uses.[25] The group noted that there is a substantial and growing interest in the use of synchrotron radiation for industrial purposes in Europe and that several synchrotron radiation sources in Europe had started to open their infrastructure to industrial users on an ad-hoc basis.[26] To the extent that new research programmes in material sciences were under way in Japan and the United States that combined the use of synchrotron radiation with industrial applications, the group recommended that synchrotron radiation sources in Europe should be made more widely available to industry. With regard to the ESRF project, the study group recommended that it should "develop an organizational structure compatible with the time scales and deadlines required in industry related research."[27]

It can be assumed that one reason for the ESRP group to work at CERN was that this location enabled the members of the project group to design and conceptualise a site-independent proposal for the ESRF because its final site was at that time yet to be decided. After its creation in the mid-1950, CERN had developed not only into a world-leading laboratory for particle physics but also into an important hub to arrange and initiate different activities related to peaceful scientific collaboration. Originally, it was located in Switzerland, which is considered neutral ground not only for politics per se but also for the politics that stand behind collaborative efforts in science. CERN was, for instance, at the very origin of collaborative activities between scientists from the East and the West during the Cold War.[28] It can further be assumed that another reason to locate the ESRP group at CERN was that this place provided a useful infrastructure (e.g. library and office spaces) and a stimulating intellectual environment. Moreover, at one point in the early history of the ESRF, the option of co-locating the ESRF with CERN was briefly discussed. This appeared as an attractive alternative from a financial point of view, as infrastructures could be shared and equipment reused. But "many of the SR [synchrotron radiation, author's note] scientists had not enjoyed the symbiosis with the high energy physics community; they did not want to be in the shadow of the 'big brother' again"[29] (see also Chap. 3).

4.2 Intergovernmental Arrangements

As mentioned in the previous section, decisions on a site and initial funding for the ESRF project required agreement and commitment on the highest political level among national governments. In 1983 and 1984, intergovernmental negotiations, mainly between delegations from France,

Germany and the United Kingdom, began to parallel the work of the above-mentioned CERN-based ESRP group (*Group Buras*). The negotiations also complemented the work of the intergovernmental progress committee (*Group Levaux*). These trilateral meetings can be regarded as a turning point for the ESRF that moved the project to the political agenda.[30]

The three countries had very different national priorities in Big Science (see Table 4.1): The British SNS[31] (Spallation Neutron Source), the German SNQ (Spallations-Neutronenquelle) and the collaborative high-flux reactor at ILL in France represented three different prioritised national projects in research with neutrons. While HERA (Hadron-Elektron-Ring-Anlage), located at DESY, was the major German project proposal for particle physics research, the collaborative LEP (Large Electron Positron Collider) at CERN was among the French and British priorities in particle physics. The proposed ESRF project as a collaborative European effort, the French SuperACO (Anneau de Collisions d'Orsay) project, the British SRS and the two German facilities HASYLAB (Hamburger Synchrotronstrahlungslabor) as part of DESY as well as BESSY (Berliner Elektronenspeicherring-Gesellschaft für Synchrotronstrahlung mbH)

Table 4.1 Portfolio of national priorities in Big Science as of 1983/1984

	Scientific field					
	Neutrons		*Synchrotron radiation*		*Particle physics*	
	Project	*Status*	*Project*	*Status*	*Project*	*Status*
France			SuperACO	Planning		
Germany	SNQ	Planning	HASYLAB BESSY	Operation (upgrade needed) operation	HERA	Planning
United kingdom	SNS	Construction	SRS	Operation		
Collaborative efforts	ILL	Operation (upgrade needed)	ESRF	Planning	LEP	Construction

Sources: French Ministry for Industry and Research, *Note sur la Concentration Tripartite Franco-Germano-Britannique en Matière de Grands Equipements*, 17 June 1983, No. 1992 0550/3, Archives Nationales de France; French Ministry for Industry and Research, *Procès-Verbal de la Réunion du 7 Septembre 1983 du Groupe Tripartite des Scenarios de Programmation des Très Grands Équipements de la Recherche Fondamentale*, 5 December 1983, No. 1992 0550/3, Archives Nationales de France.

constituted major projects in research with synchrotron radiation among the three countries.[32]

Among these different projects that were put on the agenda, two particularly mattered in the sense that they importantly shaped trilateral negotiations: The German HERA project and the British SNS project. HERA was a particle physics project proposed by the German national research centre DESY. Initial political support among German governmental representatives at the beginning of the 1980s was only lukewarm, mainly because of HERA's high costs and its risky technological design. The situation changed when the expert committee *Großprojekte der Grundlagenforschung* (also called *Pinkau Committee* after its chairman Klaus Pinkau) that had advised the German government on several Big Science projects recommended at the end of 1981 the realisation of HERA. The nomination of Heinz Riesenhuber as the new minister for research and technology in 1982 who thoroughly supported the project also played a role.[33] However, following the recommendations of the Pinkau Committee, the German government made clear that it would only be ready to support HERA under the condition that 30 to 40 per cent of the project costs would be financed by foreign partners.[34]

During intergovernmental meetings in the year 1983, Germany then tried to create a package deal which importantly shaped the early history of the ESRF: Germany proposed to join the ESRF project under the condition that France substantially contributes to HERA. However, the French delegation made clear that its participation in the HERA project could only be assured by a positive decision from Germany to join the ESRF project. During earlier meetings in 1981, the French authorities had already claimed that France would have difficulties to financially contribute to HERA and that a thorough examination of this issue could only take place when the national budget planning enters a new period from 1985 onwards.[35] At this point in late 1983, the establishment of both ESRF and HERA seemed to depend on one another with France and Germany mutually contributing to both projects.

The British delegation did not show any interest in HERA. It also left its interest in joining the ESRF very vague, but remained particularly stuck to its national SNS project. At this time in the mid-1980s, national research activities with neutrons absorbed a considerable part of the British science budget. The country did not only maintain its own national neutron source, the SNS, but also contributed to the ILL in Grenoble. The United Kingdom pursued two strategies to (financially) pave the way for British membership in the ESRF project: First, it tried to reduce British contributions to the trilateral ILL project by attracting new members. Second, it tried to attract foreign partners to the British SNS which would transform this national project into a collaborative one.[36]

This latter strategy was also called "europeanisation" or "internationalisation"[37] by the British delegation during the intergovernmental meetings, and it needs to be contextualised with the founding history of the ILL in the 1960s. The ILL was founded in 1967 by France and Germany. Originally, it was a trilateral initiative between France, Germany and the United Kingdom before the United Kingdom withdrew due to budget shortcuts.[38] France and Germany continued to realise the ILL bilaterally while the United Kingdom made plans for a competing national facility. These plans were, however, refused by the British government in 1972. At the same time in the early 1970s, the United Kingdom had re-launched its membership negotiations with the EEC after initial applications had been vetoed by French president Charles de Gaulle in 1963 and 1967. After the resignation of Charles de Gaulle in 1969 and the election of the pro-European British prime minister Edward Heath in 1970, the political climate between France and the United Kingdom improved. These political developments are important to consider because this new setting in European politics opened the door not only towards British membership in the EEC but also towards British full membership in the ILL project.[39] The United Kingdom eventually joined the ILL in 1973 as a third, equal partner which can certainly be characterised as a testament of how the politics of Big Science bear significant relevance to changing political dynamics in the EEC and the European integration process.

Eventually, British contributions to the ILL were much higher than expected. The United Kingdom had initially proposed to let its scientists work at the ILL by contributing an annual sum that equals 10 per cent of the operating costs of the reactor. But this idea was vetoed by France.[40] This 10 per cent would have meant a contribution of about 6.5 million French francs (approximately 270,000 British pounds in 1974 prices) for the year 1974.[41] Yet, negotiations turned out differently: The United Kingdom eventually contributed not only one-third of the annual operating costs (that equalled approximately 2.4 million British pounds in 1974) but also an additional sum of about 11.5 million French francs (approximately 1.2 million British pounds in 1974) annually over a period of ten years.[42] In addition, the country pledged a sum "for the joint benefit of the French-German associates" that compensated for one-third of the "assets acquired up the December 1972 from the operation funds,"[43] but no concrete amount was mentioned. These investments largely escalated beyond the originally proposed contribution of around 270,000 British pounds, and absorbed a substantial part of the British national science budget foreseen for research with neutrons. In this context, France and Germany promised that the next collaborative neutron source in Europe would be built in the United Kingdom to compensate for the country's large contribution to the ILL.[44]

This is the context in which the British vision to "europeanise" or "internationalise" its national SNS project needs to be understood, namely, to fulfil the promise to locate the next collaborative neutron source in the United Kingdom. The French and German delegations remained, however, quite reserved regarding this British initiative. Germany pursued its own national project in neutron research (SNQ) which, similar to HERA, also required foreign contributions to be realised. France considered contributing some scientific equipment to the SNS, but did not want to consider formal membership, mainly because there was no budget left for this.[45]

Summarising the above, the situation among the three partners was as follows: Germany and the United Kingdom both pursued ambitious national projects and needed to attract financial contributions from foreign partners in order to either realise HERA and SNQ or reallocate funding to support further activities, such as the British SNS. For the French delegation that promoted the ESRF as a collaborative, yet national priority, it became clear that contributions to this project from Germany and the United Kingdom would require a return of investment on both the British SNS project and the German HERA project (see Table 4.2).[46]

Table 4.2 National priorities in Big Science of Germany, France and the United Kingdom as of 1983

Priorities	*Germany*	*France*	*United Kingdom*
1	HERA	ESRF	Neutrons
2	Neutrons	HERA	ESRF
3	ESRF	Neutrons	HERA

Source: Ministère de l'Industrie et de la Recherche, *Procès-Verbal de la Réunion du 7 Septembre 1983 du Groupe Tripartite des Scenarios de Programmation des Très Grands Équipements de la Recherche Fondamentale*, 5 December 1983, No. 1992 0550/3, Archives Nationales de France

There is hence much to suggest that political rationales and logics along the lines of self-interest were among the main drivers of the trilateral intergovernmental negotiations. Of course, scientific quality and reliability of these projects constituted important aspects for the governments to legitimate large public funding and long-term commitment. However, negotiations on the three projects HERA, SNS and ESRF were also understood in terms of a tit-for-tat-strategy: investment was accompanied by a request for a return on investment. The way the tripartite negotiations proceeded resonated that it was not the idealist strive for the common good that made countries collaborate but political calculations.

4.3 Putting the ESRF in Place

As discussed in the last section, France and Germany, together with the United Kingdom, played a dominant role in the early history of the ESRF project. Germany officially announced in February 1984 that it wants to join the ESRF project. This decision came, to some extent, suddenly. The following months saw a number of crucial bilateral decisions between France and Germany that laid the foundation for the ESRF to become a reality. Most importantly, France and Germany agreed to jointly provide the majority of the construction costs for the ESRF under the condition that the facility would be built at a site of their choice.

At the same time, the *Group Buras* at CERN had prepared a site-independent proposal in 1984. The group, however, "avoided discussions on the site for the future ESRF. The site was considered a political problem and left to intergovernmental negotiations."[47] Several countries interested in hosting the ESRF project had submitted site proposals: Risø in Denmark, supported by Norway and Sweden, Trieste in Italy and Daresbury in the United Kingdom. Denmark proposed to contribute 30 per cent of the capital costs and to offer a site in Risø free of rent under the condition that the infrastructure and buildings for the project would be built by Danish companies.[48] Italy proposed to host the ESRF in Trieste and to provide 50 per cent of the construction costs, as well as additional contributions from the Italian regional government in Trieste and the government of Yugoslavia.[49] France originally nominated Strasbourg in the very early 1980s before it switched to Grenoble. Strasbourg did not only host the ESF and the French CNRS, this site proposal was also welcomed by Germany, as it "would allow German ESRF scientists to live in Germany, thereby avoiding all complicating features of moving across borders, such as tax rules, schools and other services, and general cultural barriers."[50] However, Grenoble, potentially benefiting from a co-location with the ILL, was also considered as an alternative site candidate by France. After Germany officially declared to take part in this project, the city of Dortmund and the Saar region submitted applications for hosting the ESRF. In both applications, it was argued that the ESRF would provide a truly economic boost for the two regions that were in the midst of structural change.[51]

In October 1984, the French and German ministers Laurent Fabius and Heinz Riesenhuber announced that they would jointly propose a site in Grenoble. This bilateral agreement was tied to a bilateral decision that located the ETW (European Wind Tunnel) at Cologne, Germany. The creation of a collaborative wind tunnel project had already been decided in 1983, but the question of a site was still pending: Germany preferred Cologne. France was in the favour of Toulouse, home of the Airbus company.[52] The ETW project was to become the first facility of its kind in Europe. It was of major interest for the French and German national industries, and could be considered as a competitive project with regard to the establishment of the US-American NTF (National Transonic Facility) in 1984.[53]

Although the Grenoble site was announced as a joint bilateral decision, this was certainly not the preferred location for Germany. France's original (and early) nomination of Strasbourg was clearly a strategic means to attract and secure Germany's participation in the ESRF project. However, after it became clear that the ETW project would be located in Cologne, Germany, France considered a site for the ESRF in Strasbourg (very close to the German border) too beneficial for its German partner.[54] It can only be speculated as to what extent Germany had been informed about this change of location from Strasbourg to Grenoble.

In the following months, France and Germany tried to get their site proposal accepted by other potential member countries of the ESRF. However, the way these two countries pushed their proposal through the meetings of the progress committee caused heavy contestation from the other delegations. Several European countries had proposed own site candidates (see above) and did not like that the French and German delegations portrayed Grenoble as a definite decision. The contestation of the Italian delegation during a meeting of the progress committee in late October 1984 was indicative for the general intergovernmental atmosphere at that time:

> The agreement between the German and the French delegation which is now presented for the common site of Grenoble, has been realized out of this Committee regardless of the work made in order to compare different candidatures. This kind of procedure cannot be accepted also for the reason that it gives the other countries only one alternative: to join or not the proposal.[55]

Critique also arose concerning "whether France and Germany intended to establish a truly European ESRF rather than a German French enterprise"[56] regarding the fact that the ESRF was intended to be co-located with the ILL, which was operated by France, Germany and the United Kingdom. France and Germany claimed that this project was a European one, and the French delegation argued that "the reason which had led the French government to choose the Grenoble site was precisely that it did not wish to build a German-French ESRF, but a European laboratory."[57]

As described in Chap. 2, the site selection process for a single-sited Big Science facility reaches far beyond the (simple) question of geography. The selection of a site literally puts the project in place, and is often considered as a crucial pre-condition for any other preparatory activity, such as negotiations on financial shares, or the legal and organisational framework. In other words, there was much more at stake than to simply agree on a location for the ESRF. Despite their bilateral effort on site selection, France and Germany demanded support and financial contributions from other European countries to give access to all scientists and researchers from Europe, and in order not to limit the scope of the project to a German-French context. The two countries also announced to substantially contribute to the project costs. There is thus much to suggest that this clearly put pressure on other potential member countries to either join this bilateral effort or to leave the project.

A French-Italian bilateral summit took place in early June 1985 to calm down tempers and to secure Italian commitment to the ESRF project.[58] This summit resulted in several package deals between the two countries: While Italy announced its support for a new French supercomputer project, France committed to support the creation of an Italian particle physics laboratory. It also seemed likely that France would support the establishment of a national Italian synchrotron radiation facility at Trieste, today's Elettra, which could be regarded as a compensation for the ESRF not being built at Trieste. Moreover, France lifted its veto on the location of a planned Joint Research Centre of Euratom that was to be built in Italy.

Apparently, one reason for vetoing this location was that France did not want Italy to waste money on this Euratom project but to channel efforts into the ESRF project.[59]

In parallel to the above-described political negotiations on a site for the ESRF project, numerous scientific measurements and site studies were recommended by the national delegations in the progress committee and were carried out throughout the years 1984, 1985 and 1986. These investigations should guarantee the best-informed scientific choice of a site. Geophysical and geological explorations investigated the quality of the soil. Attention was also paid on the perturbations and the vibrations caused by, for instance, railways or highways (this was called *cultural noise*), as well as the electric fields of the high-voltage lines nearby.[60] A cross-country site study in 1984 that compared the sites of Trieste in Italy, Risø in Denmark and three different sites in the Grenoble region concluded that "it was fairly certain that a suitable site could be found at *each* of the three locations."[61] However, the national delegations in the progress committee were aware that this context "was not competent to make the final steps" on the site selection, but that it "ha[s] to result from contacts between the Ministers interested in the realisation of the project."[62]

During a meeting of the progress committee in March 1985, the British delegation proposed to move the issue of site selection to the attention of the research ministers of the European member countries, which met in Rome in April 1985.[63] The Italian delegation supplemented that in this context, the ministers could "make a really European decision" that resonated the spirit of the ESRF project as a collaborative facility in Europe.[64] The Swedish and Finnish delegations supported this idea. Yet, the German delegation claimed that "the Council of EEC Ministers did not dispose of a voting and selection procedure which would solve the problem of the site of the ESRF (…)."[65] Similarly, the French delegation warned that "handing over the decision to the EEC Council of Ministers would mean denying three years of work of the progress committee and it would certainly delay the building of the ESRF."[66] On the one hand, decision-making in the Council of Ministers usually required qualified majority agreement among *all* European countries and not only among those involved in the ESRF project at that time. On the other hand, and related to this, such a procedure would also certainly run counter to the French-German bilateral efforts to define the (political) contours of the project and to locate it in Grenoble.

France put a sudden end to the concern of site selection by sending invitations to other European countries to participate in the first meeting of the provisional council of the ESRF in Grenoble which was scheduled for June 1985. The creation of a provisional council for the ESRF project, as a pre-condition to further advance this realisation of the ESRF, was proposed by the French delegation during a meeting of the progress committee in March 1985. The first meeting of the provisional council took place with delegations from Austria, Belgium, Denmark, Finland, Germany, France, the Netherlands, Norway, Spain, Sweden, Switzerland and the United Kingdom. This meeting resulted in the signing of a Memorandum of Understanding in December 1985 between the British, French, German and Italian governments.[67] Apparently, the signing of the Memorandum of Understanding put a political end to the site selection process. It located the ESRF in Grenoble but also provided funding for a preparatory phase. The costs for the preparatory phase, which was scheduled to last until the end of the year 1986, were divided between France (40 per cent), Germany (30 per cent), Italy (15 per cent) and the United Kingdom (15 per cent).

It also needs importantly to be considered that a location of the ESRF in Grenoble was not the original French site proposal. Until 1984, the country promoted Strasbourg in the region of Alsace as a French site candidate (see above). Political stakes for the Strasbourg location were high because this project was believed to significantly contribute to economic growth in the region. The ESRF as an international research infrastructure was also expected to strengthen international standing of the city of Strasbourg. The political decision in favour of Grenoble and the rejection of Strasbourg in October 1984 hence caused severe protest. Reactions culminated in the publication of a booklet in 1984 entitled *Le Livre Blanc d'un Contrat Rompu* (roughly: The White Book of a Dropped Contract),[68] as well as in a boycott of a visit of French president François Mitterrand to Strasbourg in November 1984.[69] It was, as Pierre Papon remarks, a true "psychodrame."[70]

Already in 1980, the council of the University of Strasbourg had started to draft an application to host the ESRF.[71] In a meeting in February 1981, CEA and CNRS, the major French research organisations and funding agencies, nominated Strasbourg as the French site candidate, and a commission was set up that would prepare a site proposal. The neighbourhood of the CNRS at Cronenbourg in the suburban area of Strasbourg was regarded as the most promising location.[72] However, it was also stressed during this meeting that Grenoble, as an alternative location, should be studied.[73] Pierre Papon recalls that the university in Strasbourg at that time had established a sound research community in the emerging field of structural biology that made extensive use of synchrotron radiation.[74]

A Contrat de Plan[75] concluded between the French government and the region of Alsace in 1982 mentioned the support of the French government for the location of a synchrotron radiation source in Strasbourg. This Contrat de Plan, as a newly established instrument for regional planning in France, stressed in Art. 30.5 that

> the French government will advocate the candidature of the City of Strasbourg for the hosting of new international organizations towards its European partners: Trademark Offices, the European Research Institute for Economic and Social Affairs, European Synchrotron Radiation Source, etc.[76]

The LOP (Loi d'Orientation et de Programmation pour la Recherche et le Développement Technologique de la France), a special law concerning the stipulation of national research and technological development that was passed in 1982, also mentioned the construction of a synchrotron radiation source in France. However, the law did neither explicitly mention Strasbourg nor any other definite site in France as a location for this facility. Yet, the local actors at Strasbourg, mainly politicians and scientists, wanted to see the location of Strasbourg being included in the LOP. This request can certainly be considered as a way to obtain a kind of promise at a time when the political decision on the definitive location of the ESRF was still pending.[77] Jean-Pierre Chevènement, minister for research and technology, argued in a letter to Pierre Pflimlin, mayor of Strasbourg, in 1982 that the site selection for the European synchrotron radiation source also essentially depended on the decision of the European partners, and that a more prominent status of Strasbourg as the site for the ESRF project within the framework of this law was not possible.[78]

The situation changed in 1984 when the intergovernmental negotiations among the European partners had made considerable progress and Germany had agreed to participate in the ESRF project. The new French minister for research, Laurent Fabius, stated in a letter to Marcel Rudloff, the new mayor of Strasbourg, in 1984 that the site proposal of Strasbourg only got a very hesitant recognition by the other European partners, but that the Grenoble site was quite welcomed. The United Kingdom was particularly interested in the Grenoble site under the condition that the

ESRF would be co-located with ILL. Fabius stressed that the success of the Strasbourg site depended merely on the German decision in terms of whether the country would decide to propose its own site in Germany or whether it would support the French proposal.[79]

When France and Germany announced that the final location of the ESRF would probably be in Grenoble, this certainly was a harsh setback for the actors in Strasbourg. As a compromise and compensation for the ESRF project, a Van de Graaff accelerator, the so-called *Vivitron* was later built in Strasbourg. Yet, there is also evidence, as Robert Chabbal, science administrator and general director of the CNRS in the 1970s, recalled, that the local authorities in Strasbourg acted over the course of the site selection process in a rather naïve spirit. They believed that to the extent that the French government had once decided that the synchrotron could be built in Strasbourg, this decision would hold once and for all.[80] But Chabbal stressed that decision-making on the ESRF "went beyond this simple argument. When you launch such a project, it needs to be supported by a team that believes in it, who is going to fight, who will do what has to be done."[81]

It is also true that the eventual choice of Grenoble was influenced by local politics and key actors in the Grenoble region, most importantly, Louis Mermaz, president of the Regional Council of Isère who had also been speaker of the National Assembly and entertained personal ties to the French president François Mitterrand and prime minister Laurent Fabius, as well as the mayor of Grenoble who had been chairman of the Research Commission of the Regional Council of Rhône-Alpes.[82] Moreover, Jules Horowitz, director of the Institut de Recherche Fondamentale (IRF) of the CEA

> did not like that the CEA, which was to be one of the French members of the ESRF project, should disperse its activities in too many different places. The CEA was not present in Strasbourg, but of course well-established in Grenoble.[83]

Three sites in the Grenoble region became seriously considered as locations for the ESRF project: Presqu'Ile, Sassenage and Voreppe-Moirans.[84] The Presqu'Ile (also called *peninsular site* or *Polygone Scientifique*) is located at the confluence of the rivers Drac and Isère, and already hosted several important institutions such as CNRS, CEA, INSERM, ILL, CNET as well parts of the University of Grenoble (see above). The Sassenage site of the ESRF was placed on the left side of the Drac river. The construction of a bridge was underway at that time to connect this site to the peninsular site. A third site had been made available in the industrial zone of Voreppe-Moirans. This site was located approximately ten minutes by car from the peninsular site and allowed reliable access to the two airports in the region. Two important commercial companies had already installed research and development centres there: the aluminium conglomerate Pechiney and the semi-conductor and electronics manufacturer Thompson.[85] Among the different sites in Grenoble, the provisional council of the ESRF chose a definitive site in October 1986. The council decided "that the site was to be the peninsular site of approximately 27ha adjoining the ILL (…), on the understanding that, at no cost to the ESRF, the site would be prepared as a 'green field.'"[86] The concrete site selection in the Grenoble region amended and complemented the statement issued in the 1985 Memorandum of Understanding as mentioned above. It had already located the ESRF project in Grenoble, but did not mention a concrete site.

4.4 The Role of France and Germany

As described in the last sections, the French-German partnership became decisive to make the ESRF a reality: not only did the two countries agree on major initial funding, but they also jointly proposed a site in Grenoble, and successfully pushed their vision of the ESRF through intergovernmental negotiations. This resulted in the 1985 Memorandum of Understanding, which paved the way for any other action on the project to be taken. This certainly was a remarkable effort, but it also resonated longer historical trajectories of successful collaboration between France and Germany, in both science and politics.

To the extent that the two countries had played important roles in European politics and the European integration process at that time, they were also major contributors to similar collaborative Big Science projects in Europe that were established throughout the 1960s and 1970s. When a "tidal wave"[87] of new Big Science projects rushed over Europe in the 1960s and 1970s, France and Germany were among the largest financial contributors in each of these projects: CERN, ESO, ESRO, ILL, EMBO (European Molecular Biology Organization) and EMBL, IRAM (Institut de Radioastronomie Millimétrique), ESRF and ETW—all of these projects became substantially supported by France and Germany (see Fig. 4.1).

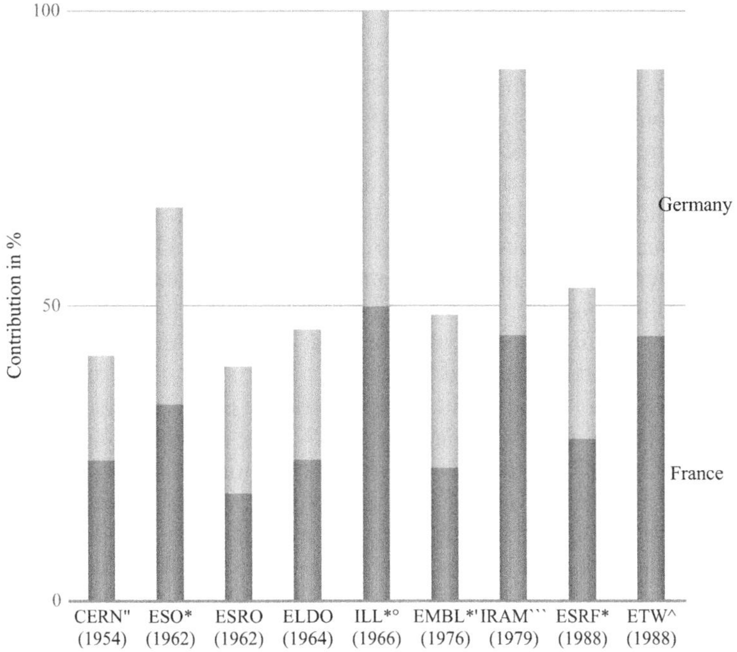

Fig. 4.1 Contributions of France and Germany in per cent (%) to collaborative Big Science projects in Europe. (Note: * as stated in the Convention; ' initial amount of funding for the year of founding; ° distribution of shares: 49 per cent (Germany) and 51 per cent (France) for the exploration phase of the reactor, during the construction and operation phase both countries contributed 50 per cent; " share until December 1956; ^ as stated in the 1985 Memorandum of Understanding. The dates in ( ) refer to the year of the signing of the Convention). (Sources: EMBL: *Scale of Contributions Calculated on the Basis of Average National Incomes, 1968–1970*, as published by the United Nations. Annex to the Agreement establishing the European Molecular Biology Laboratory (13 February 1969); ESRO and ELDO: Central Intelligence Agency (CIA). *Special Report. Western European Space Programs* (1964); ESRF: *Convention Concerning the Construction and Operation of a European Synchrotron Radiation Facility, ESRF* (16 December 1988/9 December 1991); CERN: J. Krige, "Survey of Developments." In *History of CERN: Volume I – Launching the European Organization for Nuclear Research*, eds. A. Hermann et al. (Amsterdam: North Holland, 1987), Appendix 7.1; ETW: J. Green and J. Quest, "A Short History of the European Transonic Wind Tunnel (ETW)." *Progress in Aerospace Sciences* 47, no. 5 (2011); ILL: B. Jacrot, *Des Neutrons pour la Science: Histoire de l'Institut Laue-Langevin, une Coopération Internationale Particulièrement Réussie* (Les Ulis: EDP Sciences, 2006); IRAM: P. Encrenaz et al., "Highlighting the History of French Radio Astronomy 7: The Genesis of the Institute of Radioastronomy at Millimeter Wavelengths (IRAM)." *Journal of Astronomical History and Heritage* 14, 2 (2011))

What were the main motivations for France and Germany to initiate and/or join these collaborative efforts? Certainly, the financial and energy crises in Europe, the dominant role of the United States in science and technology, as well as the strategic dimension of science and technology in the Cold War, can be regarded as some of the pieces of the puzzle to join forces in Big Science (see Chap. 1 and below).[88] However, it can also be assumed that, for France, commitment to these and similar projects was a political means to regain old strength on the European continent, and/or to shift national concerns to a European level in order to mobilise additional resources and funding.

ILL and IRAM are certainly the most remarkable efforts of what France and Germany were able to achieve in Big Science collaboration in the 1960s and 1970s.[89] Both projects constituted important historical milestones on the way towards the ESRF project, as they had demonstrated that collaborative Big Science projects between France and Germany could work effectively and successfully. The ILL in Grenoble, France was established in 1966 as a French-German bilateral effort. It became the pioneering user facility in research with neutrons on the European continent, challenging by far the capacities of running research reactors in Europe but also of the cutting-edge HFBR (High Flux Beam Reactor) at Brookhaven National Laboratory in the United States.[90] After initial participation, the United Kingdom had to withdraw from the ILL project because of budget constraints, and the French-German partnership took over the task of realising this project. The United Kingdom (re-)joined in 1973/1974 as a third partner, which shows an impressive link to European politics because the United Kingdom accessed the EEC in 1973 as a new full member (see Sect. 4.2).

IRAM was established in 1979. Initially proposed as a collaboration in radio astronomy between several European countries, it eventually resulted in a bilateral French-German effort, joined by Spain that contributed a site and office spaces. Spain eventually entered as a full member in 1990. Multilateral negotiations with other interested European countries were unsuccessful. France, which specialised in building interferometers, and

Germany, whose expertise laid in the construction of large, single dishes, decided to join forces and to converge national efforts into such a collaborative framework. The major contribution by Germany is particularly remarkable because the country had only shortly before in 1972 established the large Effelsberg Radio Telescope near Bonn.[91]

4.4.1 *"Embedded Bilateralism"*

Summarising the developments in the 1960s and 1970s, there is much to suggest that the general cooperative atmosphere between France and Germany created considerable spill-overs in science and politics contributing to deepening European political integration in the 1980s. It also mattered for the establishment of future Big Science projects, such as the ESRF, relying on the shared experience of successful cooperation in the past. In other words, the founding history of the ESRF can be regarded as a powerful testament of the strong role of France and Germany in both European politics and science at that time. In this regard, Europe provided not only a geographical but also a political frame for bilateral efforts to evolve and develop.

"Embedded bilateralism,"[92] a notion coined by the political scientists Ulrich Krotz and Joachim Schild, particularly addresses these mutual connections between the bilateral French-German relationship and the European context. Similarly, historians Corine Defrance and Ulrich Pfeil argued that the links between the French-German partnership and the European contexts also resonate how "bilateral relations are becoming more European-oriented and Europe is an opportunity to strengthen French-German cooperation and demonstrate its dynamism."[93]

Recalling the early history of the French-German partnership after the end of the Second World War, the signing of the (revised) Paris Treaties in 1954 certainly was a decisive event. This lifted not only Germany's occupation status and the ban on nuclear research. It also opened the door to German membership in the Atlantic Alliance and the defensive alliance of the Western European Union (WEU). Several years later France and Germany also found a common ground for the Saar question, compromised on the establishment of Euratom in 1957 and even agreed on joining forces for the construction of nuclear missiles.[94] The latter point was certainly a powerful sign of trust in this post-war period heavily marked by remaining mistrust.

However, this increasing political collaboration between France and Germany did not thus happen out of altruism. Following historian Keith Middlemas, it can be characterised as a marriage of convenience that was far from being an equal partnership.[95] While Germany was on its way of re-gaining economic strength at this time, France countered by its strong nuclear military force and, compared to Germany, a sounder international standing underscored, for instance, by its permanent seat in the United Nations Security Council.[96]

The Elysée Treaty, signed in 1963 by French president Charles de Gaulle and German chancellor Konrad Adenauer, can be regarded as a symbolic effort of reconciliation after the end of the Second World War and a major bilateral effort on political, scientific, cultural and military cooperation.[97] After the resignation of German chancellor Konrad Adenauer in 1963, Germany experienced changing heads of government and different political parties in power. With Ludwig Erhard (1963–1966), Kurt Georg Kiesinger (1966–1969) and Willy Brandt (1969–1974), three different chancellors governed the country within only a decade. While French president Charles de Gaulle remained in power until 1969, bilateral relations experienced a difficult and intense period, as the economic priorities of Erhard and de Gaulle did not ever go easily together.[98] De Gaulle put main emphasis on the French desired political leadership in Europe. The French veto on British membership applications in the 1960s and the Empty Chair crisis, which meant a withdrawal from all meetings of the EEC in 1965 and 1966, are only two remarkable examples of the French dominant and powerful role in Europe.[99] With the election of German chancellor Helmut Kohl and French president François Mitterrand, the 1980s became a "period of consolidation"[100] of the bilateral relations that benefited, most notably, from the amicable relationship between Kohl and Mitterrand. Sound bilateral relations also escalated to the European level, with France and Germany as driving forces in overcoming a deep crisis of the EEC in the 1980s.

The early 1980s were a difficult time for the member countries of the EEC and the pursuit of the European integration process but also a severe test for the French-German partnership. Unsolved intra-European political issues, ranging from budgetary questions to a reconsideration of the CAP (Common Agricultural Policy), loomed in parallel to serious challenges from abroad, such as the US-American SDI (Strategic Defense

Initiative) or the growing economic competition from the United States and Japan. The late 1970s saw the second oil crisis and a decrease in oil production that also caused economic recession, high inflation and unemployment in Europe in the early 1980s.

In this seemingly gridlocked situation that fundamentally halted further efforts in European integration, France and Germany decided to join forces to promote and mediate compromise between the member countries of the EEC but also to push national interests forward. Both countries took advantage of their presidency of the Council of the EEC, held by Germany in the first half of 1983, and by France in the first half of 1984, to frame and shape the European political agenda. By the rotation system of the Council of the EEC (now called Council of the EU and also known as Council of Ministers), every six months another member country took over the presidency of the council and chaired the council meetings. This also meant that:

> for a period of six months each state in turn (…) became the leading Community member both within the Community and outside it. Each of the states (…) became a leading diplomatic force in the world, able to push its own agenda, and project its own image.[101]

Ulrich Krotz and Joachim Schild, as well as historian Georges Saunier, convincingly illustrated that the years 1983 and 1984 "witnessed an intensification of bilateral contacts on all levels."[102] France and Germany agreed, for instance, on the successive suspension of controls at the French-German border, a kind of bilateral pre-agreement to the Schengen agreement in 1985. Both countries also found common ground on pending European concerns regarding the CAP and the science and technology initiative EUREKA, before these issues escalated to the wider, multilateral European context.[103] The research ministers of the two countries also officially met very frequently throughout the years 1983 and 1984.[104]

The Fontainebleau summit in June 1984, the concluding summit of the French presidency period, was decisive for the future of European politics and the success of the French-German crisis management. The former French foreign minister Roland Dumas called it the "summit of the last chance."[105] Prior to the Fontainebleau summit, the French president François Mitterrand had conducted a "highly active shuttle diplomacy,"[106] visiting the heads of governments of other European member states and developing a close personal relationship with each of them.[107]

Keith Middlemas supplements that all pending issues and open political questions "had been brought into line, largely by Mitterrand, now at his peak, yet ever conscious of the need for German backing."[108] The outcome was a complex and densely meshed net of package deals, by which compromises on common Western European concerns were concluded by satisfying, in turn, the particular national interests of the European member countries.[109]

When France held the presidency in the first half of the year 1984, responsibility for the further successful advancement of the European integration process, as well as domestic credibility, weighed hard on the shoulders of president François Mitterrand. According to Georges Saunier, to master the Community's political crisis with its head held high, France proposed to extend the policy competences of the Community:

> From the French perspective, the idea was to get out of this crisis with heads held high, that is, by advancing the Community through the adoption of new policies, particularly in the field of research and innovation.[110]

France played a pioneering role in promoting the issue of research and innovation in the European context when the French president Mitterrand came into power in 1981, and with him a leftist government. Mitterrand initiated several new policy directives for which the country would need the support of its European partners. These issues included, for instance, the reorganisation of the French public sector that went along with increasing concerns of the international competitiveness of several major industrial sectors, also affecting sensitive issues regarding price stability and payment balance, but also social and economic security.[111] Georges Saunier explained that "[t]he idea is to succeed in 'communitarising' the Community program, or, more pragmatically, to get some solidarity with, if not understanding for, the new French political orientations."[112]

France had already started to promote a stronger common European voice in research-related affairs in the early 1980s in several contexts, such as the Versailles G-7 summit. Moreover, both the report *Technologie, Emploi et Croissance* and the memorandum *Une Nouvelle Étape pour Europe: Un Espace de l'Industrie et de la Recherche* were initiated by the French government and published in the early 1980s.[113] These documents did not only testify to the strong and dominant French role in promoting a stronger European voice in research-related policy issues. But they also provided a baseline that other European countries were invited to join.[114]

Regarding more concrete measures, a general agreement on the first Framework Programme (European Framework Programme for Research and Technological Development, FP) was approved in July 1983 by the Council of Ministers, after being initiated and prepared by the European Commission. But only after several additional meetings of the council in February and June 1984 under the French presidency, a final agreement for the FP could be reached. Georges Saunier argued that this programme was very much born out of the crisis of the EEC in the early 1980s, although it also clearly testified of earlier ideas and visions.[115]

Indeed, the first FP, that ran between 1984 and 1987 marked a fundamental milestone in the strive of the EEC towards a coherent and strategic research policy. The novelty of this framework programme was that the different individual programmes "worked *together* towards the aims of Community policy in the fields of agriculture, industry, communications etc."[116] The initiatives within the first FP did not target research per se, as an overriding community goal, but rather the scientific dimension of several topical issues on the political agenda of the EEC: raw materials, energy resources, working and living conditions. It also referred to the effective use of the community's potential in science and technology, which meant, first and foremost, its exploitation in an application-oriented and industry-related context.[117]

In parallel, the European Commission launched a remarkable set of common European research-related projects: ESPRIT (European Strategic Program on Research in Information Technology) ran from 1983 onwards and BRITE (Basic Research in Industrial Technologies) reached final agreement in 1985. A preliminary definition phase of RACE (Research and Development in Advanced Communications technologies in Europe) was set up in 1985.[118] According to Georges Saunier, these initiatives were a demonstration of the Commission's capabilities and capacities to introduce new policy directions on its own. However, this ensemble of research-related projects also testified, to a certain extent, to the convergence of French and Commission-initiated interests into a common European research policy in the early 1980s.[119]

Summarising the political situation in Europe in the early-to-mid-1980s, Roland Dumas, the former French foreign minister from 1984 to 1986 and 1988 to 1993, remembered this period as follows:

> To really exist in the eyes of the world and of itself, Europe needs perspectives; it must concretely realize joint projects that make it aware of its

immense possibilities and that remove any complexity of inferiority compared to its US-American and Asian partners and competitors.[120]

Europe needed a new perspective and new projects to progress, advance and consolidate. The quotation above marks a splendid summary of how and why the issue of research entered the European arena in the 1980s. The successive introduction of research as a new policy field of the EEC certainly was a sign of its time, but it was also a topic that was expected to provide new impetus for advancing and deepening the European integration process, as well as for mastering and solving pending political conflicts. The year 1984, then, was not only of fundamental importance for the future of the EEC but also a decisive time for the ESRF project. While the European member countries, under the heading of a "European relaunch," during the Fontainebleau summit had eventually found common ground on the most urgent and conflict-ridden topics on their political agenda, this summit was also a key date to bring research on top of the European agenda, and to set political sails for the ESRF project. To the extent that France and Germany played key roles in the successful outcome of the council meeting at Fontainebleau in June 1984, the bilateral decision by France and Germany in the fall of 1984 on initial funding and a site in Grenoble for the ESRF was a similar, but common French-German demonstration of their political strength and unity in Europe at that time. Although formally disentangled from European mechanisms and official policies by evolving as an intergovernmental project in basic science, it can be argued that the ESRF provided an additional (collaborative) impetus for European integration to revive.

4.4.2 National Agendas in France and Germany

Two different national political agendas on Big Science needed to be brought together when France and Germany joined forces in the spring of 1984 to negotiate the ESRF project in a bilateral context. The motivations and rationales of the two countries to participate in the ESRF differed greatly, testifying to what historian John Krige described as "the pursuit of national self-interest by other means."[121] The joint action on a final site and initial funding was, however, also a demonstration of what these two countries were able to achieve in collaboration.

France mainly pulled the strings behind the scenes of intergovernmental negotiations on the ESRF project and had a profound interest in realising this facility on French territory. This was mainly due to the French scientific community that had raised its voice to get funding and support for a new synchrotron radiation source operating in the hard X-ray regime. France's leadership in the ESRF project benefited from its national experience with the synchrotron radiation laboratory LURE (Laboratoire pour l'Utilisation du Rayonnement Électromagnétique) and its storage rings ACO/SuperACO (Anneau de Collissions d'Orsay, Orsay Collision Ring) and DCI (Dispositif de Collission dans l'lgloo, Igloo Collision Machine), all located at Orsay near Paris.[122] LURE was created in the early 1970s. The storage rings ACO and DCI were originally built for particle physics experiments, then parasitically used by synchrotron radiation researchers, and eventually transformed into dedicated synchrotron radiation sources (see also Chap. 3). After upgrading the performances of ACO (and changing the name from ACO to SuperACO), this source started operation in the soft X-ray regime in 1984. Moreover, DCI could also operate in the hard X-ray range by pushing its parameters to limits. But there did not exist a dedicated synchrotron radiation source in the hard X-rays in the early 1980s in France.[123]

However, support for a new synchrotron radiation source in the hard X-rays was neither unanimous in French politics nor among the French scientific communities. On the one hand, close observers of this project argued that the ESRF has been regarded by French officials as another *national* milestone in its strive for European leadership in research with synchrotron radiation, challenging German national efforts in this field that put their BESSY facility in Berlin into operation in 1981.[124] Moreover, in the meantime, in 1982, a Loi d'Orientation et de Programmation (LOP) was adopted that mentioned European scientific and technological cooperation as a national priority and that aimed at facilitating and strengthening related efforts. The released funding from this law would later become crucial in the establishment of the ESRF in France.[125] Moreover, France seemed willing to realise the ESRF project also in a national context if other countries would refuse to join the French initiative.[126] On the other hand, in April 1984, agreement between CEA, CNRS and MST (Mission Scientifique et Technique, Mission on Science and Technology) could not be reached on officially granting priority to the ESRF project among other national Big Science projects.[127] Several leading and influential French scientists in those scientific fields that are directly related to research with synchrotron radiation such as biology, chemistry or microelectronics also remained opposed to increasing investments into new Big Science projects and a French contribution to the ESRF project.[128] In this context, the very nature of the ESRF as a *collaborative* effort that distributes responsibilities, duties and financial burden over many shoulders appeared as a promising strategy to the French government. In what followed in the early history of the ESRF, France balanced between two different contexts: first, the collaborative design of the ESRF, and second, the national (scientific) demand for a more powerful synchrotron radiation source in the hard X-rays.[129] Put differently, although France had promoted the ESRF as a national priority during intergovernmental negotiations with Germany and the United Kingdom, this strategy was not necessarily backed by agreement in French politics and science. But France lifted this project to the European arena by what historian John Krige characterised as "'Europeanizing' all or part of the national effort."[130] By selling it as a collaborative effort to its European partners, France strategically pursued its national Big Science priorities by avoiding the heavy financial burdens of a solo effort.

The German BMFT (Bundesministerium für Forschung und Technologie, Federal Ministry for Research and Technology) had generally supported plans to establish a collaborative synchrotron radiation source in Europe since the late 1970s. However, until the mid-1980s and, more precisely, until 1984 when Germany officially declared to join the ESRF project, research with synchrotron radiation seemingly remained of second-order importance for Germany's national research policy agenda.[131] In 1981, the national expert committee *Großprojekte der Grundlagenforschung* (also called *Pinkau Committee* after its chairman Klaus Pinkau) stated that with HASYLAB (and its use of DORIS as partly dedicated synchrotron radiation source) at DESY as well as BESSY, the German scientific communities are well equipped to pursue research with synchrotron radiation in the years to come. The committee recommended the improvement of the experimental opportunities in these places. It did not prioritise the creation of the ESRF as a collaborative effort in research with synchrotron radiation but gave way to the particle physics project HERA (see Table 4.3).[132]

Among a list of the ten Big Science projects that should be primarily supported by the German government, the ESRF only ranked eight, behind, among others, HERA, LEP and SNQ (see above).[133] By being realised in Germany, both HERA and SNQ should ensure and legitimate the future existence of two large national research centres: DESY at Hamburg where HERA should be built and KFA (Kernforschungsanstalt,

Table 4.3 Recommendations of the Pinkau Committee in 1981

Research with neutrons	*Research with heavy ions*	*Particle physics*	*Research with synchrotron radiation*
Upgrade of Berlin research reactor (BER II)	Realisation of Munich SuSe Project (superconducting cyclotron)	Realisation of LEP at CERN	Improvement of experimental capacities at HASYLAB at DESY and BESSY
Decision required on the realisation of the SNQ project	Refusal of Jülich S-JULIC project (superconducting cyclotron)	Realisation of HERA at DESY	New assessment of the ESRF project in a few years
Refusal of a project proposal for a medium flux reactor (MSR)	Decision postponed on SIS at GSI (heavy ion synchrotron)		

Note: The committee also made recommendations on a research vessel and deep drilling that are not listed above

Source: J. Rembser, "Neue Großgeräte für die Grundlagenforschung in der Bundesrepublik Deutschland." *Physikalische Blätter* 40, no. 5 (1984): 115–120

Nuclear Research Institut) at Jülich where the SNQ should be installed.[134] Moreover, Germany still heavily invested into research with neutrons and heavy ions as well as particle physics. Administrators at the BMFT argued that Germany could not afford new expensive basic research tools in the emerging area of synchrotron radiation, while still operating large facilities and projects in nuclear and particle physics.[135]

In the fall of 1982, it became clear that DORIS could not both provide sufficient experimental time for research on synchrotron radiation and respond to increasing demands by particle physicists. It can only be speculated why this new situation could not be anticipated by the *Pinkau Committee* that had issued its recommendations several months earlier.[136] The *Pinkau Committee* was then charged with the preparation of an additional assessment with a particular focus on three Big Science projects: HERA, SNQ and ESRF. The results of the second *Pinkau Committee* were officially announced in December 1983.[137] Based on these recommendations, Josef Rembser, at the time Director General for Basic Research, Research Coordination and International Cooperation at the BMFT concluded that Germany would extend the national capacities in research with synchrotron radiation at HASYLAB and BESSY but also actively participate in the planning and construction of the ESRF facility.[138]

It can only be speculated about the reasons for this sudden switch that now paved the way towards German participation in the ESRF project. The termination of the prioritised SNQ project in June 1985 probably played a role. There is much to suggest that the new Pinkau recommendations that altered Germany's position towards the ESRF as well as a limited national science budget impacted the already uncertain status of the SNQ project at that time.[139] The avoidance of a French solo effort might also have played a role. There is much to suggest that once the ETW project was brought into play, the German government perceived the package deal of locating the ETW in Germany and the ESRF in France beneficial enough to also commit to the ESRF project.[140]

4.5 Towards a Convention for the ESRF

After the Memorandum of Understanding had been signed in late 1985 by the British, French, German and Italian research ministers, further negotiations on the legal and organisational framework were handed over to the members of the national delegations that were represented in the provisional council of the ESRF project. The four initial signatories of the Memorandum of Understanding were successively joined by Spain, as well as delegations from Switzerland, Denmark, Finland, Norway and Sweden, the latter four acting together as the Nordsync consortium.[141]

The British, French, German and Italian delegations had very early agreed that the legal and organisational structure of the ILL should serve as a model for the ESRF project. Although the Italian delegation was initially opposed to this idea, it soon joined this proposal.[142] The legal basis of the ESRF project was then defined by three major texts: the intergovernmental convention to be signed on the ministerial level of the respective member countries, a protocol regulating the financial contributions, intellectual property rights and staff and procurement regulations, as well as the statutes of the company to be signed among the national proxy organisations of the member countries to create the company that would be ruled by French law.[143] To put it differently, ESRF became, similar to the ILL, a national organisation supported by international agreement.[144] The co-location with the ILL and the adoption of most of the rules and regulations of the ILL was certainly welcomed by three of the signatories of the 1985 Memorandum of Understanding because France, Germany and the United Kingdom ran ILL as a trilateral facility.

This, together with the above-discussed dominant role of France and Germany on site selection and initial funding had, however, evidently reinforced the impression among the other national delegations that the ESRF was about to become an exclusive bilateral (or at best, trilateral) project. During multilateral negotiations on the legal and organisational framework of the ESRF, the contours of British, French and German benefits and duties in the project remained a topical issue. It was generally agreed that each member would contribute at least 4 per cent of the total budget of the ESRF project. This should prevent countries to enter the collaboration at the lowest possible level, and thus to hamper the process of quickly gathering the necessary funding. This regulation was, however, not fixed in the statutes of the ESRF, but the member countries agreed on it by an exchange of letters.[145] It was also agreed that France should pay a site prime of 10 per cent on top of the construction costs to compensate for the benefits of having the ESRF located on its own national territory in Grenoble.[146]

The British contribution to the ESRF project caused controversy. In the Memorandum of Understanding, the United Kingdom had agreed to contribute 15 per cent of the costs of the preparatory phase.[147] When the agreed time range of the preparatory phase came to an end in early 1987, the United Kingdom intended to reduce its future contribution to the construction costs of the ESRF to 7 per cent.[148] This amount was later increased to 10 per cent because the British SERC (Science and Engineering Research Council) managed to make additional funds available by budget reductions in other projects and activities.[149] The British delegation argued that the step of reducing its contribution to 7 per cent became necessary because of the reluctant action from other European countries to financially support the British neutron spallation source SNS, that was renamed ISIS in 1985, which was signalled as a pre-condition for the United Kingdom to join the ESRF (see Sect. 4.2).[150]

Main opposition to the British position came from the German and French delegations. Both considered the proposed British contribution as insufficient. Yet, the other delegations could accept the eventual British proposal to contribute 10 per cent of the costs, although the Spanish delegation considered its acceptance as only provisional. A compromise on the British contribution could be found by including the British membership in the protocol, but without mentioning the exact level of its financial contribution. A footnote clarified that the British proposed a contribution of 10 per cent, and that this level was being negotiated.[151] When the convention was signed in 1988, the United Kingdom eventually contributed 12 per cent of the total costs. The amount was increased to 14 per cent when the ESRF entered its operation phase. The voting procedure constituted another controversial issue. The so-called *one country-one vote* mechanism, as adopted at CERN, was considered unfavourable. With regard to the possible future adhesion of new members to the ESRF with only a low financial contribution of 4 per cent or little more, these countries would acquire a disproportional strong weight in the voting process compared to other countries with a significantly higher contribution. Another idea, strongly promoted by France and Germany, was to adopt a voting system based on the financial shares.[152] Yet, regarding the fact that both countries, France and Germany, together held the majority of the shares, they could thus easily bypass the positions of the other member countries. A mixed voting procedure combining a majority of votes based on the *one vote-one country* system and a majority of votes based on the financial shares of each member appeared as a rather complex, but acceptable solution.[153] It was agreed according to Art. 9 of the ESRF convention that "'simple majority' means half of the capital, the number of unfavourable votes not exceeding half of the Contracting Parties," "'qualified majority' means two-thirds of the capital, the number of unfavourable votes not exceeding half of the Contracting Parties" and "'unanimity' means at least two-thirds of the capital and no counter-vote of any Contracting Party, all Contracting Parties having an opportunity to vote."[154]

A so-called *Foundation Phase Report* was published in 1987. It constituted both an interim report, summarising agreement and action taken during the preparatory phase, as well as a highlighting of the work ahead during the construction phase. Based on the report, as well as the intergovernmental agreement reached on financial contributions and the voting procedure, the governments of France, Germany, Italy, Spain, Switzerland and the United Kingdom, as well as the governments of Denmark, Finland, Norway and Sweden (acting together as the Nordsync consortium) on 22 December 1987 signed the protocol that launched the construction phase of the ESRF from January 1988 onwards. The ESRF convention and the statutes were signed in December 1988 by eleven countries: Belgium, Denmark, Finland, France, Germany, United Kingdom, Italy, Norway, Spain, Sweden and Switzerland. The Netherlands signed the convention in December 1991.

4.6 Concluding Discussion

The ESRF was established in 1988 as the first collaborative synchrotron radiation source in Europe based on an intergovernmental agreement among eleven European countries. Politics were key for the ESRF to be established and to be realised: These mattered in the intergovernmental progress committee, the ESRP, as well as tripartite negotiations among British, French and German governmental representatives.

Although formally disentangled from common European policies and integration mechanisms, the founding history of the ESRF project closely links to the strong role of the French-German partnership in both European integration and collaborative Big Science in Europe at that time. While European member countries, under the heading of a *European relaunch*, had eventually found common ground during a summit of the European Council at Fontainebleau in June 1984 on urgent and conflict-filled topics, this date was also a key to bring the issue of research (most notably through the agreement on the first FP) on top of the European agenda, and to set political sails for the ESRF project.

To the extent that France and Germany thus played a powerful role in the successful outcome of this summit, the bilateral decision by Germany and France in October 1984 on initial funding and the site for the ESRF was a similar, but usual French-German demonstration of their political strength and unity in Europe at that time. Pierre Papon argued that both countries acted as catalysers for the construction of a European space for research at the time when the competences of the EEC were only marginal in the realm of science and technology.[155] It can neither be said that only moments of strong and close political partnership between France and Germany translated into successful scientific collaboration, nor that moments of bilateral crisis resonate the absence of bilateral collaborative efforts in science and technology. While the ESRF indeed emerged during a phase of strong French-German relations, planning and preparations for the ILL took place during a time of deep political crisis between the two countries when the young European integration process experienced severe political disarrays. The bilateral French-German decision on a site in Grenoble and the initial funding of about 70 per cent of the construction costs provided a much contested, but eventually successful baseline that other countries could join. The bilateral effort was certainly a means to decide things according to their plans, but it was also, following an idea of historian Hartmut Kaelble, "the charm of the efficiency of rapid bilateral decisions."[156] The bilateral French-German decision opened the doors towards the eventual intergovernmental agreement of the ESRF among several European countries.

For the United Kingdom, the national neutron source SNS (then renamed ISIS) remained the national priority and seemingly stood in opposition to a British contribution to the ESRF project. The symbolic meaning and political weight of the British national SNS project escalated, however, far beyond the narrow time frame of the founding phase of the ESRF project, and it included longer historical trajectories of political rivalries in Western Europe. For the United Kingdom, the SNS project represented the fulfilment of a long-lasting promise: that the next collaborative neutron source would be built in the United Kingdom. The United Kingdom then wanted to lift its national SNS project to the European level by making European partners collaborate. In essence, the country thus adopted a similar strategy to its French counterpart. On the one hand, europeanising the SNS project would also have been symbolic for the political weight of the United Kingdom in Europe after becoming a full member of the EEC in 1973, and after unsuccessful applications in the 1960s that were overshadowed by the French political dominance on the continent. On the other hand, the realisation of the next collaborative neutron source project in Europe on British territory was promised to the country after it joined the bilateral French-German ILL project in Grenoble in 1973. The United Kingdom now wanted to see this promise fulfilled.

Notes

1. Germany always refers to West Germany in this context. The situation in East Germany is not considered here.
2. See ESRF, *Foundation Phase Report* (ESRF: Grenoble, 1987), ESRF Archive, 543.
3. See European Commission, *Objectives and Instruments of a Common Policy for Scientific Research and Technological Development* (Brussels: European Commission, 1972), 34–35.
4. See P. Collins, *The Royal Society and the Promotion of Science since 1960* (Cambridge: Cambridge University Press, 2016), 220.
5. See D. Greenberg and P. Boffey, "Europeans Set Up Science Foundation." *Change* 5, no. 10 (1973/1974), 78.

6. See K. C. Cramer, "Lightening Europe: Establishing the European Synchrotron Radiation Facility (ESRF)." *History and Technology* 33, no. 4 (2017), 406.
7. See Cramer, "Lightening Europe", 407.
8. See Cramer, "Lightening Europe", 407.
9. ESF, *Synchrotron Radiation: A Perspective View for Europe* (Strasbourg: European Science Foundation, 1977), ESRF Archive.
10. See ESF, *Synchrotron Radiation.*
11. ESF, *Synchrotron Radiation*, 1.
12. See K. Witte, *History of the ESRF*, manuscript of a talk given at the Science Days in Aussois, May 2001.
13. Witte, *History of the ESRF*, 3–4.
14. See ESF, *European Synchrotron Radiation Facility: The Feasibility Study* (Strasbourg: European Science Foundation, 1979), ESRF Archive.
15. ESF, *Feasibility Study*, 62.
16. See, for example, O. Hallonsten, *Big Science Transformed: Science, Politics and Organization in Europe and the United States* (Cham: Palgrave Macmillan, 2016), 140–141; O. Hallonsten and T. Heinze, "From Particle Physics to Photon Science: Multidimensional and Multi-Level Renewal at DESY and SLAC." *Science and Public Policy* 40, no. 5 (2013).
17. See Gutachterausschuss "Synchrotronstrahlung" bei der Koordinierungsstelle Hochenergiephysik, *Speicherringe für Synchrotron Strahlung. Bericht einer Studiengruppe* (1977), 1–2.
18. See Cramer, "Lightening Europe", 407.
19. See H. Schmied, "The European Synchrotron Radiation Story – Phase II." *Synchrotron Radiation News* 3, no. 6 (1990).
20. See *Minutes of the Ninth Meeting of the Progress Committee Concerned with the European Synchrotron Radiation Facility*, held in Brussels on 14 March 1985, No. 1994 0426/10, Archives Nationales de France.
21. See *Minutes of the Ninth Meeting of the Progress Committee Concerned with the European Synchrotron Radiation Facility*, held in Brussels on 14 March 1985, No. 1994 0426/10, Archives Nationales de France.
22. J. Als-Nielsen et al., *The Case for a European Synchrotron Radiation Facility* (1982), ESRF Archive, 13.
23. See Als-Nielsen et al., *The Case for a European Synchrotron Radiation Facility*, 13.
24. See *European Synchrotron Radiation Facility. Report of the ESRP*, presented by B. Buras and S. Tazzari (Geneva: CERN, October 1984), ESRF Archive.
25. See E.-E. Koch, R. Haensel. and R.H. Williams, *Industrial Uses of Synchrotron Radiation. Report of the ESRF Study Group*, October 1984.

Annex VI to: *European Synchrotron Radiation Facility. Report of the ESRP*, 4.
26. See E.-E. Koch, R. Haensel. and R.H. Williams, *Industrial Uses of Synchrotron Radiation. Report of the ESRF Study Group*, October 1984. Annex VI to: *European Synchrotron Radiation Facility. Report of the ESRP*, 4–5.
27. See E.-E. Koch, R. Haensel. and R.H. Williams, *Industrial Uses of Synchrotron Radiation. Report of the ESRF Study Group*, October 1984. Annex VI to: *European Synchrotron Radiation Facility. Report of the ESRP*, 6.
28. See, for example, J. Krige, "The Ford Foundation, European Physics and the Cold War." *Historical Studies in the Physical and Biological Sciences 29, no.* 2 (1999); B. Strasser, "The Coproduction of Neutral Science and Neutral State in Cold War Europe: Switzerland and International Scientific Cooperation, 1951–69," *Osiris* 24 (2009).
29. See Schmied, "The European Synchrotron Radiation Story II", 24.
30. See Schmied, "European Synchrotron Radiation Story II," 26; Cramer, "Lightening Europe", 408ff.
31. Documents consulted at the French National Archives often refer to the acronym SNS for *Spallation Neutron Source* whereas this name is widely unknown in the scholarly literature. The SNS early became renamed to ISIS (no acronym) and opened in 1985 under this name. This book uses both names—SNS and ISIS—interchangeably. Today's SNS (Spallation Neutron Source) at the US-American Oak Ridge National Laboratory carries the same name as the British SNS project, and should not be confused with this project.
32. See Cramer, "Lightening Europe," 408–409.
33. See Cramer, "Lightening Europe," 409.
34. So far, the usual way to realise large projects in particle physics was to build the accelerator complex by a national institute and to provide access to the instruments and the accelerator to foreign research groups through the submission of a request for experimental time. HERA was supposed to become the first project where foreign partners would also participate in the construction of the accelerator complex, while the project would remain formally connected to the national research centre DESY.
35. See *Bericht von Prof. Soergel über die Gespräche mit den ausländischen Partnern*. Annex to: Statusbericht HERA, DESY, 24 November 1983, No. 1992 0550/3, Archives Nationales de France.
36. See Cramer, "Lightening Europe," 410–411.
37. See Cramer, "Lightening Europe," 411.
38. See D. Pestre, "The Prehistory of the Franco-German Laue-Langevin Institute." In *History of European Scientific and Technological Cooperation*,

eds. J. Krige and L. Guzzetti (Luxembourg: European Communities, 1997), 142.

39. See Cramer, "Lightening Europe," 399.
40. See, for example, B. Jacrot, *Des Neutrons pour la Science: Histoire de l'Institut Laue-Langevin, une Coopération Internationale Particulièrement Réussie* (Les Ulis: EDP Sciences, 2006), 102. See also: *Lettre Envoyée par Fax en Janvier 1972 par Brian Flowers à Creyssel, Président du Comité de Direction de l'ILL.* In Jacrot, *Des Neutrons*, 157–158.
41. The background against which these sums are calculated is the limit of 65 million French Francs fixed for the operation of the research reactor in the year 1974 as stated in Art. 3b of the *Agreement Concerning the Accession of the Government of the United Kingdom of Great Britain and Northern Ireland to the Convention of 19 January 1967 as amended by the Protocol of 6 July 1971 between the Government of the French Republic and the Government of the Federal Republic of Germany on the Construction and Operation of a Very High Flux Reactor* (signed 19 July 1974).
42. Art. 2 of the *Agreement Concerning the Accession of the Government of the United Kingdom of Great Britain and Northern Ireland to the Convention of 19 January 1967 as amended by the Protocol of 6 July 1971 between the Government of the French Republic and the Government of the Federal Republic of Germany on the Construction and Operation of a Very High Flux Reactor* (signed 19 July 1974).
43. Art. 2 of the *Agreement Concerning the Accession of the Government of the United Kingdom of Great Britain and Northern Ireland to the Convention of 19 January 1967 as amended by the Protocol of 6 July 1971 between the Government of the French Republic and the Government of the Federal Republic of Germany on the Construction and Operation of a Very High Flux Reactor* (signed 19 July 1974).
44. Art. 5 of the *Agreement Concerning the Accession of the Government of the United Kingdom of Great Britain and Northern Ireland to the Convention of 19 January 1967 as amended by the Protocol of 6 July 1971 between the Government of the French Republic and the Government of the Federal Republic of Germany on the Construction and Operation of a Very High Flux* Reactor (signed 19 July 1974).
45. See Cramer, "Lightening Europe," 411.
46. See French Ministry for Industry and Research, *Note à l'Attention à Monsieur le Directeur Adjoint du Cabinet. Objet: Grand Equipment Scientifiques Européens* (Réunion de Cabinet du 2 Mars 1984), 2 March 1984, No. 1990 0594/23, Archives Nationales de France.
47. ESRF, *Foundation Phase Report*, 18.
48. See European Synchrotron Radiation Facility (ESRF)/Laboratoire Européen de Rayonnement Synchrotron, *Rapport Complémentaire au*

Document Date du 6.12.84, No. 1998 0422/1, Archives Nationales de France.

49. See European Synchrotron Radiation Facility (ESRF)/Laboratoire Européen de Rayonnement Synchrotron, *Rapport Complémentaire au Document Date du 6.12.84*, No. 1998 0422/1, Archives Nationales de France.
50. O. Hallonsten, *Small Science on Big Machines: Politics and Practices of Synchrotron Radiation Laboratories* (Lund: Research Policy Institute, 2009), 216–217.
51. See, for example, Stadt Dortmund, *ESRF in Dortmund*, ESRF Archive; *Beschreibung des Standortes Homburg (Saar)*. Bericht einer Arbeitsgruppe der Universitäten Saarbrücken und Kaiserslautern, No. 1992 0550/3, Archives Nationales de France, Paris, France; *ESRF, Standort Homburg, Universitäten in Saarbrücken und Kaiserslautern/ Lieu d'Implantation: Homburg. Universités à Sarrebruck et Kaiserslautern/ Proposed Site: Homburg. Universities in Saarbrücken and Kaiserslautern*, No. 1992 0550/3, Archives Nationales de France.
52. See Cramer, "Lightening Europe", 414.
53. See, for example, E.-H. Hirschel, H. Prem, and G. Madelung, *Aeronautical Research in Germany: From Lilienthal until Today* (Heidelberg: Springer, 2004), 402, 595; P. Papon, "European Scientific Cooperation and Research Infrastructures: Past Tendencies and Future Prospects." *Minerva* 42, no. 1 (2004), 68; J. van der Bliek, *ETW, a European Resource for the World of Aeronautics: The History of ETW in the Context of European Aeronautical Research and Development* (Cologne: ETW, 1996).
54. See *Letter from Professor Tanche, Professor Dupuy, Michel Suscillon, Daniel Bloch and André Michaudon to Luis Mermaz*, 11 October 1984, No. 1998 0422/1, Archives Nationales de France.
55. See *Approval of the Minutes of the Meeting held on the 26th of October. Comments of the Italian delegation.* Annexed to: *Minutes of the 8th Meeting of the Progress Committee concerned with the European Synchrotron Radiation Facility*, held in Brussels on 5 December 1984, B 196/146157, Bundesarchiv.
56. See *Minutes of the 8th Meeting of the Progress Committee Concerned with the European Synchrotron Radiation Facility*, held in Brussels on 5 December 1984, B 196/146157, Bundesarchiv.
57. See *Minutes of the 9th Meeting of the Progress Committee Concerned with the European Synchrotron Radiation Facility*, held in Brussels on 14 March 1985, No 1994 0426/10, Archives Nationales de France.
58. See Cramer, "Lightening Europe", 416.
59. See Cramer, "Lightening Europe", 416.

60. See CNRS, *Projet d'Implantation du Synchrotron à Grenoble.* Note Concernant les Etudes de Sols, 12 April 1985, No. 2011 1003/205–207, Archives Nationales de France.
61. See *Minutes of the Ninth Meeting of the Progress Committee Concerned with the European Synchrotron Radiation Facility*, held in Brussels on 14 March 1985, No. 1994 0426/10, Archives Nationales de France (emphasis added).
62. See *Minutes of the Ninth Meeting of the Progress Committee Concerned with the European Synchrotron Radiation Facility*, held in Brussels on 14 March 1985, No. 1994 0426/10, Archives Nationales de France.
63. See *Minutes of the Ninth Meeting of the Progress Committee Concerned with the European Synchrotron Radiation Facility*, held in Brussels on 14 March 1985, No. 1994 0426/10, Archives Nationales de France. See also: W. Weidenfeld and W. Wessels (eds.) *Jahrbuch der Europäischen Integration 1985* (Bonn: Europa Union Verlag, 1986), 459.
64. *Minutes of the Ninth Meeting of the Progress Committee Concerned with the European Synchrotron Radiation Facility*, held in Brussels on 14 March 1985, No. 1994 0426/10, Archives Nationales de France.
65. See *Minutes of the Ninth Meeting of the Progress Committee Concerned with the European Synchrotron Radiation Facility*, held in Brussels on 14 March 1985, No. 1994 0426/10, Archives Nationales de France.
66. See *Minutes of the Ninth Meeting of the Progress Committee Concerned with the European Synchrotron Radiation Facility*, held in Brussels on 14 March 1985, No. 1994 0426/10, Archives Nationales de France.
67. See *Memorandum of Understanding Concerning the Preparatory Phase of the European Synchrotron Radiation Facility* (signed at Brussels 10 December 1985), No. 1998 0422/1, Archives Nationales de France.
68. See *Le Livre Blanc d'un Contrat Rompu: Le Laboratoire Européen de Rayonnement Synchrotron. Candidature de Strasbourg*, Strasbourg, 1984, ESRF Archive. This booklet contains a collection of sources (most of them letters) that sheds light on the background of the French site selection and the sudden switch from Strasbourg to Grenoble.
69. See J.-P. Bedei, "Le Pretexte Alsacien." *l'Unité*, no. 580 (1984).
70. Y. Farge, "L'Élaboration du Projet ESRF." *Histoire de la Recherche Contemporaine* 1, no. 1 (2012), §34.
71. See *Extrait du Procès-Verbal du Conseil de l'Université Louis Pasteur de Strasbourg I*, Séance du 19 Mai 1980. In: *Le Livre Blanc.*
72. See *Letter from Chevallier to Pflimlin*, 2 April 1981. In: *Le Livre Blanc.*
73. See *Note sur le Projet de Machine Européenne à Rayonnement Synchrotron*, 5 February 1982. In *Le Livre Blanc.*
74. See Farge, "L'Élaboration du Projet", §34.

75. See Ministre Chargé du Plan et de l'Aménagement du Territoire, *Decision of the French Conseil d'Etat from 8 January 1988*, Request No. 74361.
76. Ministre Chargé du Plan et de l'Aménagement du Territoire, *Decision of the French Conseil d'Etat from 8 January 1988*, Request No. 74361. Translated from French.
77. See, for example, *Letter from Pflimlin to Chevènement*, 7 April 1982. In *Le Livre Blanc*; *Letter from Muller to Chevènement*, 26 May 1982. In *Le Livre Blanc*.
78. See *Letter from Pflimlin to Chevènement*, 7 April 1982. In *Le Livre Blanc*
79. See *Letter from Fabius to Rudloff*, 22 February 1984. In *Le Livre Blanc*.
80. See Farge, "L'Élaboration du Projet", §35, §39.
81. Farge, "L'Élaboration du Projet", §35
82. See Cramer, "Lightening Europe", 416.
83. Witte, *History of the ESRF*, 14.
84. Other sites originally considered as potential locations for the ESRF in the Grenoble area include Montbonnot, Crolles, Champ sur Drac, Echirolles and Meylen (Le Taillat).
85. See *Réponse au Questionnaire de la Direction de la Programmation et de la Provision Budgetaire sur les Sites Français envisagés pour l'ESRF* (25 June 1984/ 28 November 1984), No. 1998 0422/1, Archives Nationales de France.
86. *Minutes of the 24 October 1986 Meeting in Grenoble of the Provisional ESRF Council* (Final Version, approved in the 12 December 1986 meeting), 17 December 1986, No. 1998 0422/3, Archives Nationales de France.
87. See H. Trischler and H. Weinberger, "Engineering Europe: Big Technologies and Military Systems in the Making of 20th Century Europe." *History and Technology* 21, no. 1 (2005), 64.
88. See, for example, J. Krige, "The Birth of EMBO and the Difficult Road to EMBL." *Studies in History and Philosophy of Biological and Biomedical Sciences* 22 (2002); B. Strasser, "The Transformation of the Biological Sciences in Post-War Europe." *EMBO Reports* 4, no. 6 (2003); C. Defrance, "France-Allemagne: Une Coopération Scientifique 'Privilégiée' en Europe, de l'Immédiat Après-Guerre au Milieu des Années 1980?" In *La Guerre Froide et l'Internationalisation des Sciences: Acteurs, Réseaux et Institutions*, eds. C. Defrance and A. Kwaschik (Paris: CNRS, 2016).
89. See, for example, Defrance, "France-Allemagne", 174–175. For the role of France and Germany in the creation of the ILL see also: Hallonsten, *Big Science Transformed*, 90; Jacrot, *Des Neutrons*.
90. See, for example, H. Atkinson, "Commentary on the History of ILL and ESRF." In *History of European Scientific and Technological Cooperation*,

eds. J. Krige and L. Guzzetti (Luxembourg: European Communities, 1997); D. Pestre, "The Prehistory of the Franco-German Laue-Langevin Institute." In *History of European Scientific and Technological Cooperation*, eds. J. Krige and L. Guzzetti (Luxembourg: European Communities, 1997); G. Bacon (ed). *Fifty Years of Neutron Diffraction: The Advent of Neutron Scattering* (Bristol: Adam Hilger, 1987); Jacrot, *Des Neutrons.*

91. See, for example, M. Grewing, "Selecting and Scheduling Observations at the IRAM Observatories." In *Organizations and Strategies in Astronomy*, ed. A. Heck (Dordrecht, London: Kluwer Academic Publishers, 2007); P. Encrenaz et al., "Highlighting the History of French Radio Astronomy 7: The Genesis of the Institute of Radioastronomy at Millimeter Wavelengths (IRAM)." *Journal of Astronomical History and Heritage* 14, 2 (2011).
92. U. Krotz and J. Schild, *Shaping Europe: France, Germany, and Embedded Bilateralism from the Elysée Treaty to Twenty-First Century Politics* (Oxford: Oxford University Press, 2013).
93. C. Defrance and U. Pfeil, "Le Traité de l'Élysée et les Relations Scientifiques Franco-Allemandes." *Histoire de la Recherche Contemporaine* 2, no. 2 (2013): §14.
94. See U. Lappenküper, "On the Path to a 'Hereditary Friendship'? Franco-German Relations Since the End of the Second World War." In *History of Franco-German Relations in Europe: From "Hereditary Enemies" to Partners*, eds. C. Germond and H. Türk (New York: Palgrave Macmillan, 2008), 153.
95. See K. Middlemas, *Orchestrating Europe: The Informal Politics of the European Union 1973–1995* (London: Fontana Press, 1995), 324.
96. See P. Lellouche, "France-Allemagne: Le Double Déni." *Politique Étrangère*, no. 4 (2012), 740.
97. See C. Defrance and U. Pfeil, "Le Traité"; Defrance, "France-Allemagne", 174.
98. See, for example, Krotz and Schild, *Shaping Europe*; M.-T. Bitsch, ed., *Le Couple France-Allemagne et les Institutions Européennes: Une Postérité pour le Plan Schuman?* (Bruxelles: E. Bruylant, 2001); Lappenküper, "On the Path," 155–156.
99. See Middlemas, *Orchestrating Europe*, 52, 67.
100. P. Papon, "Intergovernmental Cooperation in the Making of European Research." In *European Science and Technology Policy: Towards Integration or Fragmentation?* eds. H. Delanghe, U. Muldur, and L. Soete (Cheltenham: Edward Elgar, 2009).
101. P. Taylor, *The European Union in the 1990s* (Oxford: Oxford University Press, 1996), 29.

102. See, for example, Krotz and Schild, *Shaping Europe*, 119; G. Saunier, "Prélude à la Relance de l'Europe. Le Couple Franco-Allemand et les Projets de Relance Communautaire Vus de l'Hexagone (1981–1985)." In *Le Couple France-Allemagne et les Institutions Européennes: Une Postérité pour le Plan Schuman?* ed. M.-T. Bitsch (Bruxelles: E. Bruylant, 2001).
103. See Middlemas, *Orchestrating Europe,* 108; Saunier, "Prélude à la Relance," 480.
104. Formal meetings between Riesenhuber and Fabius took place in May, October and November 1983, as well as May 1984, see, for example, from Bundesarchiv: B 196/121882, *Paris Mai 1983 – Treffen von Minister Riesenhuber mit Minister Fabius*; B 196/121883, *Bonn Nov. 1983 – Treffen von Minister Riesenhuber mit Minister Fabius*; B 196/121884, *Rambouillet Mai 1984 – Treffen von Minister Riesenhuber mit Minister Fabius*; B 196/122372, *Minister Riesenhuber und Minister Fabius im Okt. 1983 in Bonn.* Bilateral encounters on 7 and 24 April 1984 as well as in May, September and October 1984 were informal meetings. See, for example, French Ministry for Industry and Research, *Note pour le Ministre, Objets: Grands Équipments Scientifiques Européens*, No. 1990 059/423, Archives Nationales de France; French Ministry for Industry and Research, *Réunion ESRF au Cabinet du Ministre*, 30 January 1984, No. 1992 0550/3, Archives Nationales de France; French Ministry for Industry and Research, *Compte-Rendu de l'Entretien entre MM FABIUS et Riesenhuber (Bonn, 7 Octobre)*, 10 October 1983, No. 1992 0550/3, Archives Nationales de France.
105. R. Dumas, *Le Fil et la Pelote: Mémoires* (Paris: Plon, 1996), 330. Translated from French.
106. Krotz and Schild, *Shaping Europe*, 119.
107. See Saunier, "Prélude à la Relance," 482.
108. Middlemas, *Orchestrating Europe*, 107.
109. See Krotz and Schild, *Shaping Europe*, 119–120.
110. G. Saunier, "La Genèse du Premier Programme-Cadre Européen: Un Regard Francais (1981–1984)." In *La Construction d'un Espace Scientifique Commun? La France, la RFA et l'Europe après le "Choc du Spoutnik,"* eds. C. Defrance and U. Pfeil (Bruxelles, New York: P.I.E. Peter Lang, 2012), 91–92.
111. Saunier, "Prélude à la Relance," 469.
112. Saunier, "Prélude à la Relance," 470.
113. See J. Peterson, "Eureka: A Historical Perspective." In *History of European Scientific and Technological Cooperation*, eds. J. Krige and L. Guzzetti (Luxembourg: European Communities, 1997), 327 fn 1; Saunier, "La Genèse", 89.

114. See French Ministry for Industry and Research, *Compte-Rendu de l'Entretien entre MM FABIUS et Riesenhuber (Bonn, 7 Octobre)*, 10 October 1983, No. 1992 0550/3, Archives Nationales de France.
115. See Saunier, "Prélude à la Relance."
116. L. Guzzetti, *A Brief History of European Union Research Policy* (Luxembourg: European Communities, 1995), 83 (emphasis added).
117. See Guzzetti, *A Brief History*, 83–86.
118. See Saunier, "La Genèse," 93.
119. See Saunier, "La Genèse," 88, 93.
120. Dumas, *Le Fil*, 330. Translated from French.
121. J. Krige, "The Politics of European Scientific Collaboration." In *Companion to Science in the Twentieth Century*, eds. J. Krige and D. Pestre (London: Routledge, 2003), 914.
122. P. Papon, "La Politique Européenne du CNRS dans les Années 1980." *Histoire de la Recherche Contemporaine*, Tome I-N°1 (2012), §10.
123. See Cramer, "Lightening Europe", 412.
124. See Farge, "L'Élaboration du Projet", §6.
125. See Papon, "La Politique Européenne du CNRS."
126. See also Ministère de la Recherche et de l'Industrie/ Alain Landesman, *Compte-Rendu de la Reunion Organisee le 24 November 1982 à Daresbury sur la Collaboration Franco-Britannique en Matiere de Rayonnement Synchrotron*, No. 2000 0405/ 15, Archives Nationales de France.
127. See French Ministry for Industry and Research, *Note pour le Ministre*, 17 April 1984, No. 1990 059/423, Archives Nationales de France; French Ministry for Industry and Research, *Place des Projets de Grands Instruments Européens dans la Programmation Financière du CNRS* (12 March 1984), No. 1992 0550/3, Archives Nationales de France.
128. See French Ministry for Industry and Research, *Note d'Etape sur les Très Grands Instruments Européens (TGIE)*, 13 March 1984, No. 1994 0426/2, Archives Nationales de France.
129. See Cramer, "Lightening Europe", 413.
130. Krige, "The Politics of European Scientific Collaboration", 913.
131. See Cramer, "Lightening Europe", 413.
132. See Cramer, "Lightening Europe", 413.
133. See Schmied, "The European Synchrotron Radiation Story II", 22.
134. See Ministère de l'Industrie et de la Recherche, *Process-Verbal de la Reunion du 7 September 1983 du Groupe Tripartite des Scenarios de Programmation des Très Grands Equipments de la Recherche Fondamentale*, 5 December 1983, No. 1992 0550/3, Archives Nationales de France.

135. *Ergebnisvermerk Treffen Bundesminister von Bülow mit dem französischen Forschungs- und Technologie Minister Chevènement am 3. März 1982, Großprojekte in der Grundlagenforschung*, B 196/ 146,158, Bundesarchiv.
136. *Auszug aus dem Protokoll der 5. Sitzung des Gutachterausschusses "Synchrotronstrahlung" (GA.NG.15)*, 1 October 1982, B 196/ 146,158, Bundesarchiv.
137. A similar committee, *Sachverständigenkreis "Naturwissenschaftliche Grundlagenforschung"* (also called *Martienssen Committee*) also discussed the ESRF issue in 1984, and its decision can be seen as complementary to the recommendations of the second *Pinkau Committee.*
138. See J. Rembser, "Neue Großgeräte für die Grundlagenforschung in der Bundesrepublik Deutschland." *Physikalische Blätter* 40, no. 5 (1984), 118.
139. See Cramer, "Lightening Europe", 418.
140. See Cramer, "Lightening Europe", 418.
141. See *Note sur le Version "Juillet 1986" des Textes Définissant les Structures Légales du Futur E.S.R.F. (European Synchrotron Radiation Facility)*, No. 1998 0422/1, Archives Nationales de France.
142. See *Note sur le Version "Juillet 1986" des Textes Définissant les Structures Légales du Futur E.S.R.F. (European Synchrotron Radiation Facility)*, No. 1998 0422/1, Archives Nationales de France.
143. See *Note sur le Version "Juillet 1986" des Textes Définissant les Structures Légales du Futur E.S.R.F. (European Synchrotron Radiation Facility)*, No. 1998 0422/1, Archives Nationales de France.
144. See *Summary of the discussions in the "Subgroup on legal matters" (meeting held in Paris, July 12th)*, No. 1994 0426/10, Archives Nationales de France.
145. See French Ministre Délégué Chargé de la Recherche et de l'Enseignement Supérieure, *Note à l'Attention de Monsieur le Directeur du Cabinet. Objet: Situation des négociations portant sur les textes juridiques relatifs à l'ESRF*, 8 February 1987, No. 2000 0405/14, Archives Nationales de France.
146. See French Ministre Délégué Chargé de la Recherche et de l'Enseignement Supérieure, *Note à l'Attention de Monsieur le Directeur du Cabinet. Objet: Situation des négociations portant sur les textes juridiques relatifs à l'ESRF*, 8 February 1987, No. 2000 0405/14, Archives Nationales de France.
147. See French Ministre Délégué Chargé de la Recherche et de l'Enseignement Supérieure, *Négociation France-Britannique sur l'ESRF*, 10 February 1988, No. 2000 0405/14, Archives Nationales de France.
148. See *UK Position for the ESRF Provisional Council Meeting on 4th December 1987*, Annex to: French Ministre Délégué Chargé de la Recherche et de l'Enseignement Supérieure, *Négociation France-Britannique sur l'ESRF*, 10 February 1988, No. 2000 0405/14, Archives Nationales de France.

149. See *UK Position for the ESRF Provisional Council Meeting on 4th December 1987*, Annex to: French Ministre Délégué Chargé de la Recherche et de l'Enseignement Supérieure, *Négociation France-Britannique sur l'ESRF*, 10 February 1988, No. 2000 0405/14, Archives Nationales de France.
150. See *UK Position for the ESRF Provisional Council Meeting on 4th December 1987*, Annex to: French Ministre Délégué Chargé de la Recherche et de l'Enseignement Supérieure, *Négociation France-Britannique sur l'ESRF*, 10 February 1988, No. 2000 0405/14, Archives Nationales de France.
151. See *Compte-Rendu Succinct de la Réunion des Délégations du 22 Décembre 1987*, 31 December 1987, Annex to: French Ministre Délégué Chargé de la Recherche et de l'Enseignement Supérieure, *Négociation France-Britannique sur l'ESRF*, 10 February 1988, No. 2000 0405/14, Archives Nationales de France.
152. See French Ministre Délégué Chargé de la Recherche et de l'Enseignement Supérieure, *Note à l'Attention de Monsieur le Directeur du Cabinet. Objet: Situation des négociations portant sur les textes juridiques relatifs à l'ESRF*, 18 February 1987, No. 2000 0405/14, Archives Nationales de France.
153. See French Ministry for Research and Technology, *Compte Rendu de la Réunion du 15 Février 1986 du Conseil d'Administration Provisoire de ESRF*, 25 February 1986, No. 1998 0422/3, Archives Nationales de France.
154. *Convention Concerning the Construction and Operation of a European Synchrotron Radiation Facility, ESRF* (signed 16 December 1988/ 9 December 1991).
155. See P. Papon, "L'Espace Européen de la Recherche (1960–1985): Entre Science et Politique." In *La Construction d'un Espace Scientifique Commun? La France, la RFA et l'Europe après le "Choc du Spoutnik," eds.* C. Defrance and U. Pfeil (Bruxelles, New York: P.I.E. Peter Lang, 2012), 42.
156. H. Kaelble, *Les Relations Franco-Allemandes de 1945 à Nos Jours: Défis, Acquis, Options Nouvelles* (Ostfildern: Thorbecke, 2004), 22. Translated from French.

Bibliography

ESRF Archive, Grenoble, France

European Science Foundation. *Synchrotron Radiation: A Perspective View for Europe*. Strasbourg: European Science Foundation, 1977.
European Science Foundation. *European Synchrotron Radiation Facility: The Feasibility Study*. Strasbourg: European Science Foundation, 1979.

J. Als-Nielsen et al., The Case for a European Synchrotron Radiation Facility. 1982.
European Synchrotron Radiation Project: *European Synchrotron Radiation Facility. Report of the ESRP* (presented by B. Buras and S. Tazzari), October 1984.
Le Livre Blanc d'un Contrat Rompu. Le Laboratoire Européen de Rayonnement Synchrotron. Candidature de Strasbourg, 1984.
ESRF Foundation Phase Report. Grenoble: European Synchrotron Radiation Facility, February 1987.

ARCHIVES NATIONALES DE FRANCE, PARIS, FRANCE

Box No. 1990 0594/23: Direction Générale de la Recherche et de la Technologie
Box No. 19920550/3: Ministère ou Secrétariat d'État Chargé de la Recherche
Box No. 19940426/2: Ministère ou Secrétariat d'État Chargé de la Recherche
Box No. 19940426/10: Ministère ou Secrétariat d'État Chargé de la Recherche
Box No. 19980422/1: Ministère ou Secrétariat d'État Chargé de la Recherche
Box No. 19980422/3: Ministère ou Secrétariat d'État Chargé de la Recherche
Box No. 20000405/14: Direction Générale de la Recherche et de la Technologie
Box No. 20000405/15: Direction Générale de la Recherche et de la Technologie
Box No. 20111003/205-207: Centre National de la Recherche Scientifique

BUNDESARCHIV, KOBLENZ, GERMANY

B 196/ 146 158: European Synchrotron Radiation Facility (ESRF) 1977–1983
B 196/ 146 157: ESRF Standortfragen 28.10.83-30.4.88 Band 1
B 196/ 121882: Paris Mai 1983 - Treffen von Minister Riesenhuber mit Minister Fabius
B 196/ 121883: Bonn Nov. 1983 - Treffen von Minister Riesenhuber mit Minister Fabius
B 196/ 121884: Rambouillet Mai 1984 - Treffen von Minister Riesenhuber mit Minister Fabius
B 196/ 122372: Minister Riesenhuber und Minister Fabius im Okt. 1983 in Bonn

LEGAL/GOVERNMENTAL DOCUMENTS

Agreement establishing the European Molecular Biology Laboratory, EMBL (13 February 1969).
Agreement Concerning the Accession of the Government of the United Kingdom of Great Britain and Northern Ireland to the Convention of 19 January 1967 as amended by the Protocol of 6 July 1971 between the Government of the French Republic and the Government of the Federal Republic of Germany on

the Construction and Operation of a Very High Flux Reactor (signed 19 July 1974).

Ministre Chargé du Plan et de l'Aménagement du Territoire, *Decision of the French Conseil d'Etat from* 8 January 1988, Request No. 74361.

Convention Concerning the Construction and Operation of a European Synchrotron Radiation Facility, ESRF (signed 16 December 1988/ 9 Dyecember 1991)

Literature

Atkinson, H. "Commentary on the History of ILL and ESRF." In *History of European Scientific and Technological Cooperation.* Edited by J. Krige and L. Guzzetti, 144–153. Luxembourg: European Communities, 1997.

Bacon, G. "Introduction – The Pattern of 50 Years." In *Fifty Years of Neutron Diffraction: The Advent of Neutron Scattering.* Edited by G. Bacon, 1–10. Bristol: Adam Hilger, 1987.

Bitsch, M.-T., ed. *Le Couple France-Allemagne et les Institutions Européennes: Une Postérité pour le Plan Schuman?* Bruxelles: E. Bruylant, 2001.

Central Intelligence Agency. *Special Report. Western European Space Programs.* 1964. Available at CIA FOIA database: https://www.cia.gov/library/readingroom/document/cia-rdp7900927a004400100002-3, last accessed 4 February 2020.

Collins, P. *The Royal Society and the Promotion of Science since 1960.* Cambridge: Cambridge University Press, 2016.

Cramer, K. C. "Lightening Europe: Establishing the European Synchrotron Radiation Facility (ESRF)." *History and Technology* 33, no. 4 (2017): 396–427.

Defrance, C. "France-Allemagne: Une Coopération Scientifique 'Privilégiée' en Europe, de l'Immédiat Après-Guerre au Milieu des Années 1980?" In *La Guerre Froide et l'Internationalisation des Sciences: Acteurs, Réseaux et Institutions.* Edited by C. Defrance and A. Kwaschik, 169–186. Paris: CNRS, 2016.

Defrance, C. and Pfeil, U. "Le Traité de l'Élysée et les Relations Scientifiques Franco-Allemandes." *Histoire de la Recherche Contemporaine* 2, no. 2 (2013). Online available: https://journals.openedition.org/hrc/369?lang=en, last accessed 30 December 2019.

Dumas, R. *Le Fil et la Pelote: Mémoires.* Paris: Plon, 1996.

Encrenaz, P., Gómez-González, J., Lequeux, J., and Orchiston, W. "Highlighting the History of French Radio Astronomy 7: The Genesis of the Institute of Radioastronomy at Millimeter Wavelengths (IRAM)." *Journal of Astronomical History and Heritage* 14, 2 (2011): 83–92.

European Commission. *Objectives and Instruments of a Common Policy for Scientific Research and Technological Development*. Brussels: European Commission, 1972.

Farge, Y. "L'Élaboration du Projet ESRF." *Histoire de la Recherche Contemporaine* 1, no. 1 (2012). Online available: https://journals.openedition.org/hrc/185, last accessed 29 December 2019.

Green, J., and Quest, J. "A Short History of the European Transonic Wind Tunnel (ETW)." *Progress in Aerospace Sciences* 47, no. 5 (2011): 319–368.

Greenberg, D. and Boffey, P. "Europeans Set Up Science Foundation." *Change* 5, no. 10 (1973/1974): 78–79.

Grewing, M. "Selecting and Scheduling Observations at the IRAM Observatories." In *Organizations and Strategies in Astronomy*. Edited by A. Heck, 203–225. Dordrecht, London: Kluwer Academic Publishers, 2007.

Gutachterausschuss "Synchrotronstrahlung" bei der Koordinierungsstelle Hochenergiephysik. *Speicherringe für Synchrotron Strahlung. Bericht einer Studiengruppe*. 1977.

Guzzetti, L. *A Brief History of European Union Research Policy*. Luxembourg: European Communities, 1995.

Hallonsten, O. and Heinze, T. "From Particle Physics to Photon Science: Multidimensional and Multi-Level Renewal at DESY and SLAC." *Science and Public Policy* 40, no. 5 (2013): 591–603.

Hallonsten, O. *Big Science Transformed: Science, Politics and Organization in Europe and the United States*. Cham: Palgrave Macmillan, 2016.

Hallonsten, O. *Small Science on Big Machines: Politics and Practices of Synchrotron Radiation Laboratories*. Lund: Research Policy Institute, 2009.

Hirschel, E.-H., Prem, H. and Madelung, G. *Aeronautical Research in Germany: From Lilienthal until Today*. Heidelberg: Springer, 2004.

Jacrot, B. *Des Neutrons pour la Science: Histoire de l'Institut Laue-Langevin, une Coopération Internationale Particulièrement Réussie*. Les Ulis: EDP Sciences, 2006.

Kaelble, H. *Les Relations Franco-Allemandes de 1945 à Nos Jours: Défis, Acquis, Options Nouvelles*. Ostfildern: Thorbecke, 2004.

Krige, J. "Survey of Developments." In *History of CERN: Volume I – Launching the European Organization for Nuclear Research*. Edited by A. Hermann, L. Belloni, U. Mersists, D. Pestre, and J. Krige, 209–236. Amsterdam: North Holland, 1987.

Krige, J. "The Birth of EMBO and the Difficult Road to EMBL." *Studies in History and Philosophy of Biological and Biomedical Sciences* 22 (2002): 547–564.

Krige, J. "The Ford Foundation, European Physics and the Cold War." *Historical Studies in the Physical and Biological Sciences* 29, no. 2 (1999): 333–361.

Krige, J. "The Politics of European Scientific Collaboration." In *Companion to Science in the Twentieth Century*. Edited by J. Krige and D. Pestre, 897–918. London: Routledge, 2003.

Krotz, U. and Schild, J. *Shaping Europe: France, Germany, and Embedded Bilateralism from the Elysée Treaty to Twenty-First Century Politics*. Oxford: Oxford University Press, 2013.

Lappenküper, U. "On the Path to a "Hereditary Friendship"? Franco-German Relations Since the End of the Second World War." In *History of Franco-German Relations in Europe: From "Hereditary Enemies" to Partners*. Edited by C. Germond and H. Türk, 151–164. New York: Palgrave Macmillan, 2008.

Lellouche, P. "France-Allemagne: Le Double Déni." *Politique Étrangère*, no. 4 (2012): 739–754.

Middlemas, K. *Orchestrating Europe: The Informal Politics of the European Union 1973–1995*. London: Fontana Press, 1995.

Papon, P. "European Scientific Cooperation and Research Infrastructures: Past Tendencies and Future Prospects." *Minerva* 42, no. 1 (2004): 61–76.

Papon, P. "Intergovernmental Cooperation in the Making of European Research." In *European Science and Technology Policy: Towards Integration or Fragmentation?* Edited by H. Delanghe, U. Muldur, and L. Soete, 24–43. Cheltenham: Edward Elgar, 2009.

Papon, P. "L'Espace Européen de la Recherche (1960–1985): Entre Science et Politique." In *La Construction d'un Espace Scientifique Commun? La France, la RFA et l'Europe après le "Choc du Spoutnik"*. Edited by C. Defrance and U. Pfeil, 37–54. Bruxelles, New York: P.I.E. Peter Lang, 2012a.

Papon, P. "La Politique Européenne du CNRS dans les Années 1980." *Histoire de la Recherche Contemporaine*, Tome I-N°1 (2012b). Online available: https://journals.openedition.org/hrc/183, last accessed 10 March 2020.

Papon, P. *L'Europe de la Science et de la Technologie*. Grenoble: Press Universitaires de Grenoble (PUG), 2001.

Pestre, D. "The Prehistory of the Franco-German Laue-Langevin Institute." In *History of European Scientific and Technological Cooperation*. Edited by J. Krige and L. Guzzetti, 137–143. Luxembourg: European Communities, 1997.

Peterson, J. "Eureka: A Historical Perspective." In *History of European Scientific and Technological Cooperation*. Edited by J. Krige and L. Guzzetti, 323–345. Luxembourg: European Communities, 1997.

Rembser, J. "Neue Großgeräte für die Grundlagenforschung in der Bundesrepublik Deutschland." *Physikalische Blätter* 40, no. 5 (1984): 115–120.

Saunier, G. "La Genèse du Premier Programme-Cadre Européen: Un Regard Francais (1981–1984)." In *La Construction d'un Espace Scientifique Commun? La France, la RFA et l'Europe après le "Choc du Spoutnik"*. Edited by C. Defrance and U. Pfeil, 81–95. Bruxelles, New York: P.I.E. Peter Lang, 2012

Saunier, G. "Prélude à la Relance de l'Europe. Le Couple Franco-Allemand et les Projets de Relance Communautaire Vus de l'Hexagone (1981–1985)." In *Le Couple France-Allemagne et les Institutions Européennes: Une Postérité pour le Plan Schuman?* Edited by M.-T. Bitsch, 463–485. Bruxelles: E. Bruylant, 2001.

Schmied, H. "The European Synchrotron Radiation Story – Phase II." *Synchrotron Radiation News* 3, no. 6 (1990): 22–26.

Strasser, B. "The Coproduction of Neutral Science and Neutral State in Cold War Europe: Switzerland and International Scientific Cooperation, 1951–69," *Osiris* 24 (2009): 165–187.

Strasser, B. "The Transformation of the Biological Sciences in Post-War Europe." *EMBO Reports* 4, no. 6 (2003): 540–543.

Taylor, P. *The European Union in the 1990s.* Oxford: Oxford University Press, 1996.

Trischler, H. and Weinberger, H. "Engineering Europe: Big Technologies and Military Systems in the Making of 20th Century Europe." *History and Technology* 21, no. 1 (2005): 49–83.

van der Bliek, J. *ETW, a European Resource for the World of Aeronautics: The History of ETW in the Context of European Aeronautical Research and Development.* Cologne: ETW, 1996.

Weidenfeld, W. and Wessels, W., eds. *Jahrbuch der Europäischen Integration 1985.* Bonn: Europa Union Verlag, 1986.

Witte, K. *History of the ESRF*, manuscript of a talk given at the Science Days in Aussois, May 2001.

CHAPTER 5

Establishing the European X-Ray Free-Electron Laser (European XFEL), 1992–2009

The European XFEL (X-ray Free Electron Laser) is a recent collaborative effort in Europe for the realisation of a free-electron laser for hard X-rays. It was established as a limited liability company (Gesellschaft mit beschränkter Haftung, GmbH) under German domestic law based on an intergovernmental agreement that was signed in 2009 by twelve countries: Denmark, France, Germany, Greece, Hungary, Italy, Poland, Russia, Slovakia, Spain, Sweden and Switzerland. It opened to external users in September 2017. The free-electron laser is based on a linear accelerator and on various undulators, all accommodated in an underground tunnel system, leading over a total length of 3.4 km from the German research centre DESY (Deutsches Elektronen-Synchrotron) in Hamburg-Bahrenfeld to the town of Schenefeld in the federal state of Schleswig-Holstein.

DESY played a key role in managing and organising various kinds of preparatory activities of the European XFEL project, including the initiation of a public plan approval procedure or the supply of office spaces and computational infrastructure. With the establishment of the European XFEL GmbH in 2009 through intergovernmental agreement, DESY became Germany's representative and shareholder in the project. DESY and the European XFEL collaborate on the basis of a long-term agreement.[1] The history of DESY and its gradual transformation over the course of the last decades, from a mission-oriented particle physics laboratory to a multidisciplinary user facility with a main focus on photon science, are

K. C. Cramer, *A Political History of Big Science*,
Palgrave Studies in the History of Science and Technology,
https://doi.org/10.1007/978-3-030-50049-8_5

important aspects against which the founding history of the European XFEL needs to be set and understood. The first section of this chapter is hence dedicated to this important aspect relying on previous research on the history and organisational transformation of DESY.[2]

5.1 The Transformation of DESY

The German national research centre DESY played a key role in the founding phase of the European XFEL project. It provided not only the geographical context but also the institutional and organisational setting in which the European XFEL project took shape. DESY was established in 1959 as a national research centre for particle physics. Over the last decades, the research centre gradually transformed from a single-mission particle physics research centre into a multidisciplinary and multipurpose facility for particle physics, accelerator development and photon science, which importantly includes research with synchrotron radiation. Major particle physics accelerators at DESY such as DORIS (Doppel-Ring-Speicher, Double-Ring Storage), PETRA (Positron-Elektron Tandem Ring Anlage, Positron-Electron Tandem Ring Accelerator) and HERA (Hadron-Elektron-Ring-Anlage, Hadron Electron Ring Facility) became successively upgraded and dedicated to research with synchrotron radiation (DORIS III and PETRA III) or were shut down (HERA).[3]

At the time of the establishment of DESY in the late 1950s, the political climate in West Germany and Western Europe was favourable to propose a large and single-mission facility for particle physics. By the mid-1950s, the ban of the Allied forces on nuclear research in West Germany was lifted and the newly founded Federal Ministry for Atomic Issues (Bundesministerium für Atomfragen) possessed enough resources to fund large and cost-intensive research projects. Physics and engineering sciences had acquired a prestigious and powerful political status in the post-war and early Cold War years throughout Western Europe and the United States. Generous funding for particle physics and political commitment to collaboration among European countries had made the creation of CERN (European Organization for Nuclear Research) in 1954 possible.[4] Through massive investments into ever-larger and complex accelerator-based particle physics experiments, the United States had become the spearhead of particle physics research by the 1960s, and the point of reference for European countries to economically and scientifically catch up and compete with (see Chap. 1).[5]

The first particle accelerator at DESY, an electron synchrotron, was based on an idea of DESY's first director Willibald Jentschke.[6] This accelerator was proposed as a complementary effort to the plans of CERN for a proton synchrotron. Lifting DESY's envisaged scientific programme to the same level as CERN, at least rhetorically, put an ambitious agenda in place, namely, that DESY was expected to compete with CERN and to develop into a renowned particle physics facility on its own. Scientists from DESY and CERN had jointly discussed their plans, for instance, during a conference on high-energy particle accelerators at CERN in 1956, which provided a forum to plan and coordinate major accelerator-based projects in Europe. The electron synchrotron DESY, which also gave name to the entire research centre, went into operation in 1964.[7] It was decided that DESY would become the first research facility in West Germany that would not only provide experimental opportunities for a single university-related institute but that it should be open to all scientists and researchers in West Germany. This provided an excellent opportunity for scientists in Germany to make use of cutting-edge experimental equipment and to establish a competitive standing compared to the United States, as well as compared to other Western European countries that had recently launched national particle physics projects or that were about to do so.[8] Electron synchrotrons, similar to that of DESY but with a considerable lower energy, were planned and went into operation in the 1960s in Liverpool (United Kingdom), Bonn (West Germany), Lund (Sweden) and Frascati (Italy).[9]

In the following years, the portfolio of DESY was expanded by constructing new machines that improved, amended and/or replaced former projects. The underlying storage ring technology of DORIS, built between 1969 and 1975, was still in its infancy when the final decision on the project was taken in the late 1960s (see Chap. 3). Only small prototypes and limited technological know-how existed for this new kind of accelerators. The construction of PETRA, built between 1975 and 1978, was a similarly remarkable effort in completing a major accelerator project below budget and ahead of time of a competitive project in the United States, namely the PEP (Positron-Electron Project) at SLAC (Stanford Linear Accelerator Center).[10] As mentioned above, an underlying principle of DESY was to work complementary to the activities of CERN. In the late 1970s, CERN had proposed the LEP (Large Electron-Positron Collider), a large circular-shaped accelerator that would collide electrons and positrons. DESY followed with the proposition of a complementary effort, namely the circular-shaped accelerator HERA, which would collide electrons and protons. HERA went into operation in 1990. The first experiments were conducted in 1992.[11] HERA was very much based on the initiative of particle physicist Björn Wiik. In the 1990s, he became the director of DESY and emerged as one of the leading figures in the TESLA project before his sudden death in 1999. As will be shown later, the TESLA project can be considered as one of the very birth places of the European XFEL, and Wiik also played a decisive role during the project's early history.[12]

HERA introduced a new way of establishing and organising Big Science projects. The so-called *HERA model* constituted an innovative funding scheme for the construction of large accelerator complexes. So far, national institutes had constructed and operated large accelerators on their own budget. External research groups had gained access through the submission of applications and a successful peer-review process. External groups also participated in the construction of the detectors and similar experimental equipment, but construction work on the accelerators remained in the hands of the national institutes. However, for HERA, substantial foreign contributions were also required for the construction of the accelerator complex. Otherwise, the German government had argued that it would not provide the remaining and necessary funding (see also Sect. 4.2).[13] The HERA model was innovative in the sense that DESY remained a national facility. But the way foreign partners contributed to the HERA project was similar to the collaborative and intergovernmental framework of other Big Science facilities in Europe such as CERN or ILL (Institut Laue-Langevin). While DESY and the German government, as the main funders, remained in full control of the project, the need for foreign contributions made HERA a crucial element of multilateral bargaining and negotiating compared to other collaborative and intergovernmental projects (see also Sect. 2.1.3). The basic principles of the HERA model, namely, to fund a project through international collaborative efforts that remained located at a national facility, also mattered for the founding phase of the European XFEL (see below).

In scientific and technological terms, HERA became the first occasion for DESY to use superconducting magnets for a newly built accelerator complex. Superconducting magnets can produce very intense magnetic fields while being cooled down to very low temperatures. This was a new and cutting-edge approach for the magnetic lattice of accelerators, and it was expected to challenge by far the performance parameters of non-superconducting designs.[14] Pioneering efforts with superconducting technology were carried out by the US-American national laboratories Brookhaven and Fermilab (short for Fermi National Accelerator Laboratory) in the 1970s. This principle was also applied to the US-American SSC (Superconducting Supercollider) as well as (later) for the TESLA linear collider and free-electron laser proposal. The free-electron laser of today's European XFEL project became also based on superconducting cavities, which makes the performance of the facility outstanding compared to other free-electron lasers (see Chap. 3). This superconducting technology is also the agreed basic technology for the ILC (International Linear Collider), which is expected to become the next-generation project of the international particle physics community and, simply speaking, promoted as an expanded and global version of the TESLA linear collider.[15]

Next to its expertise in particle physics, DESY successively developed a second field of activity and competence: research with synchrotron radiation. By the early 1960s, research with synchrotron radiation emerged as a side activity at the particle physics accelerators at DESY.[16] However, to the extent that synchrotron radiation was originally an unwanted by-product of particle physics experiments, research with synchrotron radiation thus depended on the goodwill of particle physicists to share their equipment and to provide beam time (see Chap. 3). Although this was also the general situation at DESY, research with synchrotron radiation soon developed into an increasingly demanded experimental resource at the research centre, and the number of users grew constantly throughout the 1960s and 1970s.[17] DORIS became partly used as a synchrotron radiation source. But the once high hopes of the researchers on synchrotron radiation soon turned into disappointment, mainly because of beam instabilities and a change of operation mode induced by particle physicists.[18] In the early 1970s, the EMBL (European Molecular Biology Laboratory) established an outstation at DESY to closely connect on a long-term basis to the experimental opportunities in research with synchrotron radiation. At the end of the 1970s, DESY decided to establish HASYLAB (Hamburger Synchrotronstrahlungslabor, Hamburg Synchrotron Radiation Laboratory), a laboratory entirely dedicated to research with synchrotron radiation, which coordinated all synchrotron radiation activities at DORIS.[19] In the 1980s, a growing number of applications were submitted to use experimental opportunities at DESY for research with synchrotron radiation and the spending of the facility on this matter increased.[20] In the early 1990s, research with synchrotron radiation became recognised as a formal organisational goal within the DESY statutes. In 1993, DORIS became a dedicated synchrotron radiation source, while PETRA could also be partly used as a synchrotron radiation source from the mid-1990s onwards. In 2000, the research division of DESY was split into particle physics and photon science, with two research directors: Robert Klanner for particle physics and Jochen Schneider for photon science, who was also the head of HASYLAB. In 2001, the German government prioritised the conversion of PETRA into a cutting-edge synchrotron radiation source (PETRA III), over the continuation of the particle physics programme of HERA, following the recommendation of the DESY directorate.[21]

When HERA was eventually closed in 2007, it was the first time that DESY shut down any of its large instruments without reusing it in new experimental contexts. Today, no large accelerator project for particle physics remains at DESY, but the particle physics division of the facility still analyses data from experiments at HERA, and also contributes to several particle physics experiments, for instance, at the LHC (Large Hadron Collider) at CERN, the SuperKEKB (KEK-B-factory) accelerator at KEK (High Energy Accelerator Research Organisation) in Japan, as well as to a neutrino experiment at the South Pole (IceCube).[22] The future of particle physics at DESY remains unclear. But there is much to suggest that the eventual shut-down of HERA in 2007 and the political commitments to PETRA III and (later) the European XFEL project during the 2000s set sails for the scientific future of DESY that apparently lies in photon science.[23] In other words, as will be illustrated in the following sections, increasing scientific and political commitment to research with synchrotron radiation at DESY that was seemingly paralleled by diminishing activities in particle physics can be interpreted as crucial pre-conditions to pave the way for the European XFEL project to come into being.

5.2 The TESLA Proposal for a Linear Collider

As described in the last section, the early 1990s constituted an important turning point for the scientific mission of DESY. Research with synchrotron radiation became recognised as a formal organisational goal, and the storage ring DORIS became a dedicated synchrotron radiation source. However, around the same time, major efforts were also put into planning a new project in particle physics. The international TESLA collaboration, located at DESY, proposed a linear collider project based on superconducting technology for particle physics research. In 1992, the TESLA collaboration submitted an initial proposal for the realisation of a test bed for superconducting cavities needed for the realisation of the linear collider. This test bed was named TESLA Test Facility (TTF).[24] This was a major cornerstone in the founding history of the European XFEL because, importantly, it was later expanded by the additional integration of a free-electron laser.

DESY had already started to participate in the development of new electron-positron linear collider designs several years ago by pursuing two different approaches; first, further development of already existing designs, based on non-superconducting s-band technology by the SBLC (S-Band Linear Collider) collaboration, and second, investigation of the new superconducting technology by TESLA collaboration.[25] The underlying linear design was a novelty for DESY, as the laboratory had never before planned, constructed or operated a linear collider. A linear collider required very different scientific and technological approaches and methods compared to circular-shaped accelerators.[26] The s-band technology was already in use at some other linear accelerator projects such as the 50-GeV linac at SLAC. It was appreciated as a sufficiently researched approach, and was generally regarded as a cost-efficient accelerator design. The TESLA project, in contrast, was a pioneering effort in using superconducting technology for a linear accelerator design.[27] As mentioned above, DESY had already implemented superconducting technology at the HERA accelerator, but a successful implementation into a linear collider design was expected to provide unprecedented experimental opportunities. This approach also challenged on-going developments at the sibling laboratories of DESY, namely, KEK and SLAC, that remained stuck to non-superconducting technology.[28] In 1998, DESY eventually decided to abandon work on the s-band project proposal to fully concentrate on the TESLA approach.[29]

After the WR (Wissenschaftlicher Rat, Scientific Council) and the EWR (Erweiterter Wissenschaftlicher Rat, Extended Scientific Council), two of the main internal governing bodies at DESY, were briefed in March 1991 about the planning of a linear collider, in June the same year, the EWR recommended to further pursue this approach.[30] After approval of the proposal to construct a prototype linear accelerator, the TTF, the TESLA collaboration started the construction of the test facility in 1993. The BMFT (Bundesministerium für Forschung und Technologie, Federal Ministry for Research and Technology) was also informed about the plans of DESY, and briefed the Committee on Research of the German Federal Parliament in March 1992.[31] As the final linear collider of TESLA would (in its initial version) encompass an underground tunnel with a linear accelerator of about 33 km in total travelling, from DESY in Hamburg north to Westerhorn in Schleswig-Holstein, the tunnel had to pass through fifteen municipalities. It also needed the commitment of the local administrations.[32] The Senate of Hamburg, the Cabinet of Schleswig-Holstein and the County of Pinneberg were introduced to the linear collider plans of DESY in January and April 1994. Other main actors, such as the prime minister of Schleswig-Holstein Heide Simonis and the mayor of Pinneberg were informed personally over the course of the year 1996.[33]

A working group with the task to create a state contract (Staatsvertrag) between Hamburg and Schleswig-Holstein was established in December 1996.[34] The state contract was signed in 1998. It did not only provide the legal basis for the future process of the project that meant, in particular, the initiation of the mandatory public plan approval procedure (Planfeststellungsverfahren).[35] But it also made DESY the main actor behind the project to promote it among international partners. Shortly after signing, DESY and the TESLA collaboration initiated a public plan approval procedure and an environmental assessment. Both measures were scheduled to be conducted in 2002 when the TESLA collaboration expected to have concluded the planning phase and had submitted the project to the German Science Council (short for German Council of Science and Humanities, Wissenschaftsrat) for evaluation.[36]

The TESLA proposal for a linear collider needs to be seen in the context of developments in particle physics in the 1980s and 1990s. This development most notably consisted of activities at CERN as well as preparations in the United States to realise the large SSC project. The TESLA proposal built on international agreement among particle physicists that had emerged in the early 1990s to promote linear electron-positron colliders as promising new research opportunities. Over the course of the 1970s and 1980s, experiments at existing circular-shaped electron-positron storage rings had already revealed exciting insights.[37] Such experiments were carried out at SPEAR (Stanford Positron Electron Asymmetric Rings) and PEP at SLAC, DORIS and PETRA at DESY, TRISTAN (Transposable Ring Intersecting Storage Accelerator in Nippon) at KEK and LEP at CERN. But further investigations with electrons and positrons required energies beyond the energy of 200 GeV that had been achieved by CERN's LEP at the end of the 1980s. Yet, *circular* designs of electron-positron colliders were not appropriate to reach higher electron energies because electrons radiate large amounts of energy when accelerated in circular paths (see Chap. 3). There was wide consensus among the scientific community that only *linear* colliders in which electrons and positrons are accelerated in opposite directions and made to collide reach energies substantially above the energy of 200 GeV of the LEP.[38] In the early 1990s, CERN was also about to plan a next large accelerator project, the LHC. The LHC, however, was conceptualised as a hadron accelerator in which protons are made to collide. With their complicated structure, protons are more difficult to control, whereas, in contrast, the collision processes at electron-positron colliders can be calculated more precisely, and the properties of new particles can be determined with rather high precision.[39] The main motivation of the LHC was to record the Higgs particle (that the LHC achieved in 2012) as one of the missing pieces in the particle physics Standard Model. In contrast, future electron-positron colliders, such as the proposed TESLA linear collider, were expected to further investigate the Higgs mechanism once the Higgs would have been recorded at the LHC. In other words, particularly when properties of newly discovered particles should be investigated and explored, experiments had shown that hadron accelerators were not best-suited for such tasks, but that complementary efforts at electron-positron colliders were needed.[40]

It was in this context that three laboratories, DESY in Germany, KEK in Japan and SLAC in the United States, that all had considerable expertise in electron-positron accelerators, signed an interlaboratory Memorandum of Understanding in 1993 in order to strengthen collaborative research and development efforts towards next-generation linear electron-positron colliders in particle physics. Other institutions and governments were invited to join the three signatories.[41] The Memorandum of Understanding was signed the same year as the SSC project in the United States was cancelled by the Congress of the United States. The termination of the SSC project provides another important point of reference for understanding the early history of the TESLA linear collider project; particularly so if compared to the LHC project at CERN. The SSC project originated in the early 1980s as a major US-American effort in particle physics. It responded but also intended to challenge recent developments in Europe, most notably at CERN where intensive work on three accelerator projects, namely, SPS (Super Proton Synchrotron), LEP and LHC, unfolded at that time.[42] The LHC was CERN's largest project, but its realisation lay still ahead in the future. The SSC outnumbered the LHC efforts of CERN by far with a circumference of its accelerator of 81 km (compared to 27 km of the LHC) and an investment volume eventually exceeding 10 billion US dollars, which tripled the envisaged costs of the LHC and was expected to be sustained by one single country, namely the United States.[43]

The SSC project was cancelled in 1993 by a decision of the US-American congress that not only shocked US-American scientists and physicists but that also questioned further work on the LHC project of CERN: "If the Americans gave up why do we [member countries at CERN, author's note] have to spend money to do this?"[44] The cancellation of the SSC bore relation to (scientific and political) mismanagement, a lack of foreign contributions and cost overruns.[45] However, it was also a strong sign for a "doctrinal shift"[46] in science policy at the end of the Cold War in the United States and Western Europe: It had gradually ended monetary generosity for very cost-intensive particle physics projects and switched priorities of research policies towards the solution of societal concerns (such as health, climate change or energy supply) beyond the disciplinary context of particle physics (see Chaps. 1 and 3).[47] The Memorandum of Understanding for a linear collider in particle physics hence needs to be seen and interpreted as an alternative route for future research in particle physics by joining forces on the development and realisation of future electron-positron colliders.[48]

5.3 From the Free-Electron Laser at the TESLA Test Facility to FLASH

Going back to the year 1993, particle physicist Björn Wiik became the director of DESY. To the extent that he had already played a key role in the HERA project, the TESLA linear collider was another project that very much relied on his ideas and enthusiasm for research at the frontier of particle physics. During his inaugural visit at the BMFT (Federal Ministry for Research and Technology) in 1993, Wiik addressed potential future projects at DESY since the large HERA project at DESY had shortly before started operation. Wiik was well aware that political and financial commitment from the BMBF to HERA was a major proof of trust; unlikely to be repeated at this rate and at this time. In this regard, it can be interpreted as a carefully considered strategy that Wiik did not explicitly request funding for a fully-fledged TESLA linear collider project, although preparations for this project were well under way at DESY. But he framed his request for additional funding and political commitment along the lines of a technology programme as a test bed for key components of the future TESLA linear collider.[49] Minister Heinz Riesenhuber and other governmental representatives seemed to be quite interested in the development of key technologies rather than constructing another major machine for particle physics research.[50] Jochen Schneider, director of HASYLAB in the 1990s, remembered the situation as follows: "[T]here were clear indications from the funding agencies in Germany that such a large research facility [the TESLA linear collider, author's note] should be attractive not just to the particle physics community."[51] Wiik's proposal eventually manifested in the realisation of the TTF.

This situation fit within the context of changing science policy rationales in the early 1990s. With the end of the Cold War, the pre-eminence in science and technology as a crucial national strategic advantage for the security and power outreach of a country in the former tension-laden bipolar atmosphere was no longer a valid reason to fund costly Big Science. Large particle physics experiments had lost much of its political prestige and political symbolism.[52] Big Science projects became increasingly placed within new settings that point to a more strategic role of knowledge, science and research for and within economy and society, and to their potential usefulness for creating commercial applications and for solving grand challenges in the late twentieth century and the early twenty-first century, such as health, climate change, ageing or energy security.[53] These developments translated into a re-direction of funding priorities and rationales for the support of, and commitment to, Big Science. Heads in the ministries and governments in Europe and the United States were increasingly turned towards application-oriented and multipurpose projects in the life sciences, material sciences and related fields, for which they needed experimental equipment and opportunities of synchrotron radiation sources, free-electron lasers and neutrons. In other words, to the extent that political framework conditions in and geopolitical contours of Europe changed at the end of the Cold War, so did the politics and organisation of Big Science. However, it is also true that continuity had its role to play. Big Science in the post-Cold War should not be regarded as "something entirely new and discontinuous but as something partly new and partly built out of existing elements and within existing institutional frameworks."[54] New (geo)political alliances, funding principles and science policy rationales should not be interpreted as a complete replacement of former priorities and settings. But, as argued by sociologist Olof Hallonsten, "[t]he transformation of Big Science shows a symbiosis between the growth of neutron scattering, synchrotron radiation, and free electron laser as well as a change in the political landscape."[55] This particularly matters with regard to the support of particle physics after the end of the Cold War, the apparent decline of which is only relative. Investments into particle physics remain large compared to other disciplines and scientific fields.[56] Relating to the situation at DESY in the early 1990s, there is hence much to suggest that the proposed technology-oriented programme to test key components for future projects at DESY aligned quite well to new political contexts.

Another import aspect was the reunification process of Germany in the early 1990s, which also marked an important turning point for the science systems of both West and East Germany. Five new federal states of the former East Germany joined the West German federal political system, and major scientific institutions from the former East were, based on the recommendations of the German Science Council, integrated into the West German scientific landscape.[57] To master the unification absorbed much political and administrative energy, as well as the financial resources devoted to it. The situation of the Big Science facilities in Germany remained, however, relatively stable, considering the fact that none of the existing projects were closed, but three new research centres in former East Germany were created. Yet, the creation of these three new facilities was not matched with a larger budget, which meant that the large-scale research facilities in former West Germany suffered severe budget cuts.[58]

Under the (political) conditions outlined in the previous two paragraphs, the future for photon science at DESY seemingly looked much brighter than that of particle physics. This seemingly was also recognised by particle physicists and researchers on synchrotron radiation at DESY that decided to forge an alliance. To the initial TESLA project proposal for a linear collider in particle physics, they added a next-generation effort in photon science: a free-electron laser.[59] The TESLA linear collider/free-electron laser alliance was represented in person by the particle physicist and director Björn Wiik and the photon scientists and HASYLAB directors Jochen Schneider and Gerhard Materlik. The two fields "were portrayed as close allies"[60] to attract political commitment to the free-electron laser but also, so it was believed by the leaders of DESY, to eventually secure funding for *both* efforts.

From the 1980s onwards, linear accelerators with laser-like characteristics[61] were expected to perform as exciting next-generation light sources (see Chap. 3).[62] Free-electron lasers became widely discussed during international workshops and conferences over the course of the 1990s. In a European context, negotiations were conducted in the context of the European Round Table for Synchrotron Radiation Research and Free Electron Lasers or the European Large Facility Group, the latter as an informal discussion forum between representatives from France, Germany, Italy, Spain and the United Kingdom. The assessments of these forums converged on the point that smaller free-electron laser facilities that operate in the IR to the VUV and soft X-ray regime should be funded by national governments.[63] Related research activities should then be coordinated within one or several network(s) funded by the EU (European Union) or similar schemes that are to be supported by the FP (Framework Programmes, short for Framework Programmes for Research and Technological Development) of the EU (see Sect. 2.2). In the spectral range of the hard X-rays, it was agreed among researchers and policymakers that only one large facility should be realised in Europe based on an intergovernmental agreement and should be funded by several countries.[64] This latter idea eventually manifested in the DESY/TESLA free-electron laser proposal.

The technical design of the TESLA linear collider was recognised to be suitable to integrate and host an additional free-electron laser.[65] For DESY, the specific choice of an X-ray free-electron laser was also motivated by its close relationship with its sibling research centre SLAC that had launched a major effort in constructing a free-electron laser in the early 1990s. During a stay at SLAC in 1992, Björn Wiik was briefed by Herman Winick, at this time deputy director at the SSRL (Stanford Synchrotron Radiation Lightsource Division) at SLAC, about the activities of the laboratory in constructing a free-electron laser, which later became the LCLS (Linac Coherent Light Source), the first free-electron laser user facility operating in the hard X-ray range.[66] It was the former HASYLAB director Gerhard Materlik who, after a sabbatical at SLAC between 1993 and 1994, eventually "convinced Wiik to include an X-ray laser in the superconducting linear accelerator project at DESY."[67]

In June 1994, the EWR at DESY was briefed about the set-up of an additional working group investigating the possibility to expand the design of the TESLA Test Facility for the linear collider to include a free-electron laser.[68] The TTF for the linear collider was considered as an ideal driver for the free-electron laser and the optimal place to test its innovative design. As part of the TESLA Test Facility for the linear collider, this test bed for a free-electron laser became known as the TTF FEL (Free-Electron Laser at the TESLA Test Facility). Although the long-term goal was to construct a free-electron laser in the hard X-ray regime, it seemed too ambitious to directly realise these final parameters. As mentioned above, DESY had never before planned, constructed or operated a linear collider, neither in particle physics nor in photon science. Moreover, the SASE mechanism, a fundamental requirement to make free-electron laser designs based on a linear accelerator and superconducting technology, in short, wavelength regions running, relied so far only on theoretical considerations, predictions and simulations for shorter wavelengths, and had only been tested experimentally at larger wavelengths in the microwave regime (see Chap. 3).[69]

DESY decided to realise a free-electron laser through a two-phased approach. First, lasing down to the VUV range that approximately corresponds to a wavelength range of 10 to 200 nm should be achieved (TTF FEL Phase I). Second, the installation should be upgraded by achieving lasing in a wavelength range of about 2 to 200 nm (TTF FEL Phase II).[70] A conceptual design report for the realisation of a VUV free electron was published in 1995, which provided a detailed science case and technological design for the TTF FEL, as well as a time schedule for the work on the TTF FEL to begin.[71] Several uncertainties that made the free-electron laser project ambitious but also risky were laid out in the conceptual design report and legitimated a careful two-phase approach. Next to the experimental testing of the SASE principle for shorter wavelengths, new designs and materials for monochromators and mirrors needed to be developed to resist the radiation damage and the heat load of the very intense beam.[72] Not only DESY but several other research groups around the world had put considerable effort into the successful experimental accomplishment of the SASE mechanism during the late 1990s. In February 2000, the proof of principle of the SASE mechanism in a wavelength range of 109 nm was successfully completed at DESY.[73] After lasing down to 98 nm was achieved in 2001 at DESY, by simultaneously lifting the performance parameters of the TTF FEL far beyond those of existing and comparable facilities, Phase I of the TTF FEL was successfully concluded. Phase II was approached by upgrading and expanding the performances of the free-electron laser.

There is much to suggest that the successful conclusion of Phase I, and the resulting unprecedented opportunities, caught the researchers at DESY with true astonishment.[74] On the one hand, the results of Phase I of the TTF FEL were integrated into a technical design report that eventually considered the combined realisation of a linear collider and a free-electron laser that should be realised in the hard X-ray regime. This eventually resulted in the European XFEL project. On the other hand, an expanded version of the free-electron laser of the TTF FEL developed into a self-standing project, becoming the first free-electron laser user facility in the world in the extreme ultraviolet and soft X-ray regime.[75] This project was named FLASH. It was commissioned in 2004, and the first users were welcomed in August 2005. DESY's successful experimental proof of the SASE principle in the short wavelength regions and ongoing developments in the United States, where SLAC thoroughly prepared for the construction of a free-electron laser in the hard X-ray regime, seemingly constituted an important trigger for the directorate of DESY. It switched efforts and manpower from the free-electron laser test facility, which was closely linked to the TESLA linear collider efforts, to the fully-fledged next-generation user facility FLASH.[76]

These developments are remarkable in two regards. First, it should be borne in mind that, originally, the test facility was conceptualised as a test bed for the key technologies of the future linear collider in particle physics. It also developed into a test bed for a free-electron laser that led to both FLASH and today's European XFEL project. Second, it can certainly be argued that the symbolic of the new name of FLASH, independent from the initial test facility, should not be underestimated. It was of major importance for both the high standing of photon science activities at DESY, and the consequences it had for the original TESLA linear collider effort. The dedication of the free-electron laser test facility to a user facility meant a thorough separation of this project from the broader and encompassing linear collider proposal of which it so far had only been a side aspect.

Eventually, the efforts of the TESLA collaboration on both the linear collider and the free-electron laser converged into a combined Technical Design Report, the TESLA TDR, published in 2001.[77] Such a double-track approach was clearly born out of strategic considerations to simultaneously meet political expectations on the free-electron laser and to keep DESY a key player in particle physics research. On the one hand, the German government showed more interest in funding an application-oriented and multipurpose photon science machine than another major piece in particle physics. On the other hand, the preference of the government of the free-electron laser over the linear collider did not necessarily resonate the genuine interests of scientists and the hopes and desires they put into realising new projects. There is much to suggest that, from the point of view of DESY and the TESLA collaboration, a free-electron laser should, scientifically and technologically, remain a side aspect of the linear collider. But it needed to be promoted as a flagship project to catch the commitment of the government to fund the whole effort.

5.4 Political Commitment to the European XFEL

When the 2001 TESLA TDR eventually considered the construction and implementation of a combined linear collider/free-electron laser facility, as described in the previous section, this report was then submitted to the German Science Council to get a positive evaluation and, consequently, the support and necessary funding of the German government for the entire project. While the linear collider certainly was the favourite project of DESY and TESLA, and the free-electron laser was considered only as a strategic and politically desired side aspect, the 2003 decision of the German government changed these plans drastically. The government preferred the realisation of the free-electron laser, whereas the linear collider effort was put to a halt.

Closely connected to this governmental decision was also the go-ahead for the realisation of FAIR (Facility for Antiproton and Ion Research), an antiproton and ion accelerator at Darmstadt, connected to the GSI (GSI Helmholtzzentrum für Schwerionenforschung, GSI Helmholtz Centre for Heavy Ion Research). The main purpose of FAIR was to improve basic research in the fields of nuclear, hadron, atomic and plasma physics, as well as application-oriented research in the areas of material science, biophysics and radiation biology.[78] To the extent that experiments with ions can be regarded as complementary to research with synchrotron radiation to investigate and explore the structure, processes and interactions of matter, the experimental tools which FAIR would provide, can be regarded as complementary to that of the European XFEL (see Chap. 3). The decision of the German government to realise these two facilities, but to put the TESLA linear collider project to a halt, certainly needs to be interpreted in this context. With the European XFEL and the FAIR facility that together required an investment of about 1.3 billion euros at the time of their proposal, Germany had hence committed to two large and cost-intensive projects for the coming decades. Foreign contributions to these projects were expected (and probably also needed), and the German government made foreign contributions from European and international partners a pre-condition to support each of the projects with the majority of the construction costs.[79]

Recalling the decision-making process for the European XFEL in the BMBF (Bundesministerium für Bildung und Forschung, Federal Ministry for Education and Research), in two letters dating March and October 2000, minister Edelgard Bulmahn had approached the German Science Council to comment on a list of nine Big Science projects, each with an investment volume of more than 15 million euros, and to assess their potential for realisation and successful operation. The submitted list comprised nine projects that were at that time under planning and/or future consideration at member institutes of the large science organisations in Germany, including the Helmholtz Association, the Leibniz Association and the Max Planck Society (see Table 5.1).[80]

The German Science Council was created in 1957 jointly by the German federal government and the governments of the federal states. It became an advisory body to assess, recommend and review projects, activities and developments that affect the German higher education and research landscape.[81] While between the 1970s and 1990s, small ad-hoc expert-based committees[82] were responsible for preparing recommendations for the German government on national and collaborative Big Science projects, the involvement of the Science Council was a novelty for the assessment of the European XFEL project and similarly large project proposals. The involvement of the Science Council in the decision-making process on large-scale research projects should reflect the importance and impact of these projects on the entire German science system and research landscape.[83]

The Science Council passed its view in July 2002.[84] With regard to the combined TESLA project proposal, conditional support was recommended for both parts: the linear collider as well as the free-electron laser. However, it was proposed to thoroughly separate the linear collider and the free-electron laser, and to initially aim for the realisation of the free-electron laser. The Science Council argued that the linear collider would probably need a global effort, and that a decision on its realisation and potential construction at DESY should be postponed until the project acquired a more formal base.[85] With regard to the recommendations of the Science Council on other Big Science projects under review, the FAIR

Table 5.1 List of nine large-scale facilities as submitted to the German Science Council in 2001 for evaluation

Name	*Applicant*	*Required investment in million euros (approx.)*	*Recommendation*
High Field Laboratory Dresden (HLD)	Helmholtz Research Centre Dresden-Rossendorf and Leibniz Institute for Solid State and Materials Research	25	Unconditional support
High Magnetic Field Facility for Neutron Scattering Research	Berlin Neutron Scattering Centre (BENSC) at the Hahn-Meitner-Institute (HMI)	49	No decision taken
High Altitude and Long Range Research Aircraft (HALO)	German Aerospace Center	97	Unconditional support
Soft X-ray Free Electron Laser (Soft X-ray-FEL)	BESSY	148	No decision taken
European Drilling Research Icebreaker (Aurora Borealis)	Alfred Wegener Institute for Polar and Marine Research (AWI)	250	No decision taken
TESLA X-ray Free Electron Laser (TESLA X-FEL)	DESY	673	Conditional support
International Accelerator Facility for Beams of Ions and Antiprotons (FAIR)	GSI Heavy Ion Research Centre	675	Conditional support
European Spallation Source (ESS)	International collaboration with major involvements of the Hahn-Meitner-Institute (HMI) and the Forschungszentrum Jülich	1390	No decision taken
TeV-Energy Superconducting Linear Accelerator (TESLA)	DESY	3450	Conditional support

Note: Grouped by amount of envisaged investment
Source: Wissenschaftsrat, *Stellungnahme zu neun Großgeräten der naturwissenschaftlichen Grundlagenforschung und zur Weiterentwicklung der Investitionsplanung von Großgeräten / Statement on Nine Large-Scale Facilities for Basic Scientific Research and on the Development of Investment Planning for Large-Scale Facilities* (Köln: Wissenschaftsrat, 2003), 106ff, 135

project also became conditionally supported. No concrete decision was taken for the Soft X-ray FEL (Soft X-ray Free-Electron Laser), the ESS (European Spallation Source), the High Magnetic Field Facility for Neutron Scattering Research and the European Drilling Research Icebreaker (Aurora Borealis). The Science Council recommended unconditional support for the HDL (High Field Laboratory Dresden) project and the HALO (High Altitude and Long Range Research Aircraft) project (see Table 5.1). As these latter proposals were less cost-intensive than TESLA and FAIR, it can be assumed that a final decision was easier to take for HALO and HLD.

Based on the recommendations of the Science Council for the TESLA linear collider and the free-electron laser, the BMBF asked DESY and the TESLA collaboration for an alternative design of the combined linear collider/free-electron laser effort. In October 2002, a new technical design report for a dedicated X-ray free-electron laser with an accelerator in a thoroughly separate tunnel was published as an amendment to the initial technical design report of TESLA in 2001.[86] In February 2003, the new project proposal of the European XFEL got the go-ahead of the BMBF, and it was decided that Germany would cover around half of the construction costs. Simultaneously, the BMBF also gave green light to the FAIR project, which had so far only conditionally been recommended by the Science Council (see above).[87]

The governmental decision on the European XFEL project can also be interpreted as a way to guide and channel future developments at DESY. Given that DESY constituted the initial institutional context of the European XFEL project, there is much to suggest that the decision of the government in favour of the European XFEL also translated back into the strategic orientation of DESY and its future scientific activities as one of the research centres of the Helmholtz Association (Hermann von Helmholtz-Gemeinschaft). In other words, "[t]he XFEL decision concluded developments at DESY, signalling that its future would be accelerator-based photon science."[88]

Moreover, it can be argued that DESY and the European XFEL both constituted visible crystallisation points for Germany's science policy strategies and ambitions in the late twentieth and early twenty-first century. To gain a nuanced understanding of the position and role of DESY within this development, the argument needs to be set in a broader context comprising longer historical developments in the German non-university research sector, such as the establishment of the Helmholtz Association, its governance structures and main funding schemes. Public funding of DESY at its time of establishment in the late 1950s was a major source of conflict between the federal government and the federal states because, in principle, the German Basic Law (Grundgesetz) from 1949 secured the independence of the federal states from the government in educational and cultural matters (cultural sovereignty, Kulturhoheit).[89]

In the case of DESY, governmental representatives claimed that the federal states should fund the project substantially. Otherwise, they would deny their principle of cultural sovereignty, which they had always guarded so carefully.[90] The representatives of the federal states argued that it should be the task of the German government to fund the DESY project because it was conceptualised as a facility for *all* scientists in Germany.[91] Eventually, it was agreed that funding was to be split between the government (85 per cent) and the states (15 per cent). This was mainly due to the expected large costs, importance of DESY for the whole German scientific community and its political symbolism that escalated beyond the regional border of the federal state of Hamburg.

More clarity on the general issue of funding Big Science was reached several years later through the introduction of Article 91b to the Basic Law in 1969 that stated that the government and states together may finance those large research facilities that are of transregional importance. The conclusion of a framework agreement (Rahmenvereinbarung) several years later between the government and the federal states required the government to finance 90 per cent of the costs and the federal states to finance the remaining 10 per cent.[92]

With the conclusion of this 90/10 funding scheme, Big Science projects became an official part of the portfolio of the German federal research ministry. However, despite the principle of cultural sovereignty, the government had started to already expand its science policy competences and capacities since the mid-1950s into fields where the federal states did not play a dominant role so far. This was, first and foremost, in the cost-intensive Big Science projects but also in applied research in the fields of military strategic research, aerospace research and research on nuclear issues. Until today, the 90/10 funding scheme remains the principle for the realisation of national Big Science facilities grouped under the Helmholtz Association, and the government thus remains the main funding body.[93]

In 1995, several national Big Science facilities[94] became grouped under the Helmholtz Association that replaced the former AGF

(Arbeitsgemeinschaft Großforschungseinrichtungen), founded in the 1970s. In 2001, the Helmholtz Association established a comprehensive internal governance structure that until today serves as an umbrella organisation for all Big Science facilities in Germany.[95] Based on the recommendations of the Science Council, a comprehensive evaluation process of the German science system and the government-funded research organisations was carried out in the mid to late 1990s. As part of this process, it was decided for the facilities (and their projects) that are grouped under the Helmholtz Association to link the allocation and distribution of funding to a new, competitive programme-oriented funding scheme.[96] However, the government remained the main sponsor providing 90 per cent of the costs. Within the internal organisational structures of the Helmholtz Association, the government is represented in the Committee of Financing Partners (Ausschuss der Zuwendungsgeber) as well as the Senate. The Committee of Financing Partners is composed of representatives from the government and the federal states that contribute to the funding of a specific Big Science facility within the Helmholtz Association. The Senate as the main governing body is composed of representatives from both scientific and political institutions in a way that the number of scientific representatives outnumbers those of the political representatives. The Committee of Financing Partners as well as the Senate are the main channels through which the research objectives and policy directions for the Helmholtz Association are set and under which funding is distributed and grouped.[97] While the government has a say in priority-setting and agenda-making of the Helmholtz Association, by being present in these governing bodies, this influence does not necessarily translate into influence on the scientific programmes and activities of the individual facilities. The individual research facilities still guard a certain autonomy regarding their research activities and scientific programmes.[98]

Summarising the above said, the direct influence of the government on the scientific programme and research activities of DESY is, in principle, limited. However, as outlined in the previous section, administrators at the ministry and governmental representatives indeed had a vivid interest in shaping the scientific agenda and future research activities at DESY. It can be assumed that the ministry and the government wanted to get a foot into the door of DESY to further advance and/or complete the facility's gradual transformation from a single-mission particle physics laboratory towards a multipurpose facility in the field of photon science, which appeared, in terms of application-orientation, societal awareness and technology transfer, much more promising than particle physics. The eventual governmental decision on the European XFEL (and also the refusal of the TESLA linear collider that the Science Council had conditionally supported) can hence be interpreted as an important strategic move within the broader national science policy context to guide and channel activities at DESY towards photon science.[99]

Before the decision of the German government on the European XFEL was made official, state secretary Wolf-Dieter Dudenhausen from the BMBF together with Hermann Schunck, general director at the BMBF, visited DESY to brief the directorate. Hermann Schunck recalls that "the atmosphere in the directorate was delicate, because of the cancellation of the TESLA project [the linear collider, author's note] and it was hardly brightened by the (conditional) commitment to the XFEL."[100] There are several reasons to suggest that the rejection of the TESLA linear collider proposal was indeed a nightmare for DESY. First, the separation of the linear collider and the free-electron laser, as required by the BMBF after the assessment of the Science Council, had already been against the initial goals and intentions of the strategic alliance between particle physicists and photon scientists at DESY.[101] Second, there is also much to suggest that the rejection of the TESLA linear collider by the ministry and government came unexpected. As described above, the state contract between Hamburg and Schleswig-Holstein that had set the legal basis for any further development in the project was already concluded in 1998, and a mandatory public plan approval procedure was initiated in 2001. These timely efforts can be interpreted as a way to speed up the preparatory process of the linear collider project prior to the construction phase. The construction phase was expected to start immediately after the project would have gotten the official go-ahead and the necessary funding by the German government in 2003 (see the first section of this chapter).

The overall future of the linear collider was indeed uncertain. It was briefly discussed among the stakeholders at the BMBF to cancel any further support of the TESLA linear collider project, but this idea was discarded relatively soon. Instead, the linear collider was not formally cancelled but "approved without providing funding."[102] DESY could hence in principle continue to participate in the development of a next-generation linear collider, promote its TESLA approach and keep its scientists and technicians involved in any future activities.[103] But the BMBF and the German government would not provide funding to do so. However, DESY kept efforts for the construction of a next-generation linear collider project high. In the following years, the innovative project design of TESLA was presented in different international forums and became thoroughly discussed by the particle physics community. One fundamental milestone for the future of the original TESLA linear collider proposal was achieved in 2004: Particle physicists agreed that their major next project, the ILC, should be based on the superconducting technology as originally proposed by the TESLA collaboration.[104] This not only provided DESY and TESLA collaboration with a key position within this project but, even more important, also mattered for today's European XFEL collaboration. The free-electron laser of the European XFEL project also became based on superconducting technology. The work necessary to make superconducting technology work for the free-electron laser can also serve as a way to test the future technology of the linear collider and to gather the necessary know-how for future projects in particle physics to come.[105]

After the decision of the BMBF on the European XFEL was made official, state secretary Dudenhausen called the representatives of the two host states, state secretary Roland Salchow from Hamburg and state secretary Hellmut Körner from Schleswig-Holstein, to inform them that both states were expected to together contribute 10 per cent of the construction costs of the European XFEL project. It was agreed that Hamburg would contribute 7 per cent and that Schleswig-Holstein would participate with 3 per cent, corresponding to their (expected) duties in and benefits from the project.[106] On 28 September 2004, a new state contract between Hamburg and Schleswig-Holstein was signed. This time, it initiated the public plan approval procedure for the European XFEL project.[107]

5.5 Foreign Partners and In-Kind Contributions

After the European XFEL project got political commitment and initial funding from the German government in 2003, the German minister Edelgard Bulmahn sent letters to several governments inviting them to join the project.[108] As the German government had decided to simultaneously support both the European XFEL and the FAIR project, foreign contributions were expected to secure the necessary funding to construct and operate these two large facilities.

In September 2004, a Memorandum of Understanding was signed by eight countries: France, Germany, Greece, Italy, Spain, Sweden, Switzerland and the United Kingdom. By December 2005, five additional partners had joined: China, Denmark, Hungary, Poland and Russia. The Slovak Republic decided to participate in 2007.[109] An International Steering Committee composed of representatives from the thirteen signatories of the Memorandum of Understanding was created to set the necessary organisational, legal and scientific framework to prepare the construction phase and to embed this project in a broader European framework.[110] On 5 June 2007, the construction of the free-electron laser began as a "start version" with less experimental capacities than initially envisaged.[111] At the end of September 2008, the "quasi final form"[112] of the legal texts were approved. However, negotiations with foreign partners were complicated and lengthy and the convention could not be concluded earlier than in November 2009. The negotiated shares of the construction costs turned out to be very unequally distributed. While Germany holds 53.6 per cent, Russia is the second biggest shareholder with 23.1 per cent and the other shareholders contribute between 0.37 per cent (Greece) and 3.33 per cent (France) of the construction costs (see Fig. 5.1).[113] This unequal distribution has been a major source of conflict and delay for the whole project.

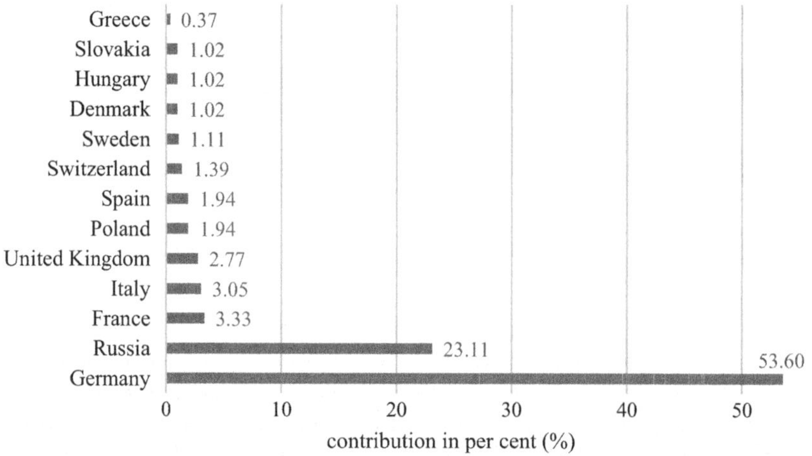

Fig. 5.1 Financial Contributions of the member countries to the construction costs of the European XFEL in per cent. (Source: Art. 5 of the *Convention concerning the Construction and Operation of a European X-Ray Free-Electron Laser Facility* (30 November 2009))

Throughout the second half of the twentieth century, France and Germany have often played major (financial) roles in collaborative Big Science projects in Europe (see also Chap. 4). Drawing on earlier collaborative experiences with France (e.g. at the ESRF), administrators at the German BMBF "thought it to be self-evident"[114] that France would also substantially contribute to the European XFEL project. But things turned out differently: France eventually contributed only 3.3 per cent of the construction costs; and this not in cash but only in kind (see Sect. 2.1.3).

In the summer of 2003, a first meeting between representatives from the BMBF led by Hermann Schunck, at that time general director of the BMBF, and the French team around Bernard Bigot, at that time vice minister Claudie Haigneré's chief of staff at the Ministry of Research and New Technologies, took place in Paris for a discussion of a potential French membership in the European XFEL project. During this meeting, the two sides agreed on a French contribution of approximately 10 per cent of the construction costs. In turn, Hermann Schunck (re-)affirmed that the German government would support the ITER (International Thermonuclear Experimental Reactor) project.[115]

ITER is a large plasma fusion project supported at this time by China, Euratom, Japan, Russia, South Korea and the United States. Preliminary discussions between France and Germany on the European XFEL project in 2003 had coincided with the site selection process for ITER. During the internal European decision-making process led by Euratom, the French site proposal for ITER in Cadarache eventually won the support of Euratom against the Spanish site proposal of Vandellos. This choice did not please Spanish officials, who portrayed it "as being imposed by a 'French-German axis.'"[116]

The (informal) agreement between France and Germany regarding a French contribution to the European XFEL was confirmed by a letter signed on the ministerial level shortly after the meeting in Paris. When Bigot very soon after the meeting changed his professional position from the ministry to the French CEA (Commissariat à l'Énergie Atomique), the agreement was suddenly revoked.[117] Moreover, a sceptical attitude prevailed among several French scientists regarding the feasibility and usefulness of free-electron lasers in the hard X-ray regime. Already in 1996, during a workshop at the ESRF in Grenoble, French scientists showed scepticism towards the realisation and technical feasibility of the then-planned LCLS in the United States.[118] The influential scientist and former director general of the ESRF, Yves Petroff, was also reserved regarding the feasibility of the technical design of the European XFEL. These aspects might also have impacted the decisions of the French governmental representatives.[119]

The United Kingdom signed the European XFEL Memorandum of Understanding in 2004 and the "minimal planned contribution" of the country to the construction costs of the project was estimated to 30 million British pounds over a five-year period.[120] Yet, in December 2009, several days before the convention was to be signed, the country announced its complete withdrawal from the project. The British STFC (Science and Technology Facilities Council) had experienced a dramatic funding shortfall in the fiscal years 2008/09 and 2009/10. The budget for the coming years thus necessitated a re-prioritisation of investments as well as significant budgets cuts to again arrive at a balanced budget. This radical step resulted in the sudden withdrawal from several national as well as international projects.[121] The financial crisis in 2008/09 and the fall in the value of the pound also probably played a fundamental role for this step to be taken.

Another critical aspect in this situation was the internal organisational structure of the STFC. The STFC is a governmental agency created in 2007 by merging the former national PPARC (Particle Physics and Astronomy Research Council), CCLRC (Council for the Central Laboratory of the Research Councils) and EPSRC (Engineering and Physical Sciences Research Council). STFC managed, among other tasks, national Big Science projects and British participation in international, collaborative Big Science projects.[122] The global budget of the agency, from which national as well as international commitments needed to be fulfilled, caused, according to British science minister Paul Drayson, uneasy tensions within the ability of the agency to meet its duties.

> [I]t has become clear to me that there are real tensions in having international science projects, large scientific facilities and UK grant giving roles within a single Research Council. It leads to grants being squeezed by increases in costs of the large international projects which are not solely within their control.[123]

To counter this vital situation at the STFC, a prioritisation process ran from May to December 2009 that ranked major national and international projects according to their scientific excellence, economic and societal impact, leadership and synergies. It resulted in a fundamental prioritisation programme that included, among other dramatic cuts, the withdrawal from the European XFEL project. Withdrawal or mission out was also announced for the Photon Science Institute and the national NLS (New Light Source) project, the ALICE (A Large Ion Collider Experiment) experiment at the LHC at CERN, as well as several projects in astronomy, such as the Cassini probe, the Venus Express Orbiter and the SOHO (Solar and Heliospheric Observatory) mission.[124]

Reconciling from above, on the one hand, drop-out from the European XFEL project hence deemed unavoidable because resources were put into already existing national and international projects and facilities. The focus was on operating national facilities, such as the national synchrotron radiation source, the Diamond Light Source, and the neutron and muon source ISIS. On the other hand, withdrawal did not signify that British researchers would be left without access to any state-of-the-art free-electron laser project. They could gain access to several free-electron laser facilities such as the LCLS at SLAC in the United States through a successful and peer-reviewed application for experimental time.[125]

Despite the formal withdrawal from the project, British scientists and engineers continued to contribute in kind to the European XFEL implementing, for instance, a pixel detector (Large Pixel Detector, LPD) and a high-energy laser.[126] The BBSRC (British Biotechnology and Biological Sciences Research Council), the MRC (Medical Research Council) and the Wellcome Trust invested 64 million British pounds into the European XFEL project over a period of five years (2014–2019), in order for British scientists to gain access to the crystallography techniques at the facility at a time when a definitive (political) decision on re-joining the project was uncertain.[127]

The withdrawal of the United Kingdom came together with a lowering of contributions from Spain and Italy that were also hit by the economic crisis. The project was left with a funding shortfall of 150 million Euro. Russia, that was also hit by the financial crisis in 2008 and 2009 as much as other countries in Europe, partly pledged the gap left by the withdrawal of the United Kingdom, and the reduction of contributions for Spain and Italy, by contributing additional 50 million Euros.[128] In 2014, the United Kingdom announced that it would consider re-joining the European XFEL project. The country eventually became the twelfth member of the project in March 2018. The United Kingdom contributes 26 million Euros, which corresponds to 2 per cent of the construction costs, to compensate for other members' investments in the construction of the

European XFEL over the past years. The country also contributes 2 per cent of the operation costs annually.[129]

Germany also considered the United States as a potential partner in the European XFEL project. But John Marburger, at that time the science advisor to the US-American president, argued during a meeting with German minister Edelgard Bulmahn that he considered an X-ray free-electron laser as a strategic project that should be pursued under own (national) responsibility.[130] According to Hermann Schunck, at that time general director at the BMBF, the United States did not want to miss the opportunity to pioneer in the use of an X-ray free-electron laser. This endeavour was characterised by the LCLS at SLAC whose construction was expected to be completed earlier than the European XFEL, and to provide unprecedented experimental opportunities.[131] Nevertheless, an interlaboratory Memorandum of Understanding between DESY and SLAC was already concluded in 2002 that improved the collaboration between the two research centres with regard to the planned LCLS and the free-electron laser of the TESLA collaboration that is now the European XFEL.[132]

China also signed the Memorandum of Understanding and Chinese representatives regularly participated in the European XFEL steering committee meetings during the first years. It can only be speculated why they did not sign the convention and left the question of their membership pending. One potential reason could be that they wanted to realise a free-electron laser by themselves, lifting their scientific capacities and capabilities to the same level as Europe, Japan and the United States. Another reason could be that it was politically overly optimistic to invite both Russian and Chinese representatives and to make them collaborate.[133]

Several of the European XFEL member countries decided to contribute in kind to the project in the form of pre-manufactured goods, equipment or staff (see Sect. 2.1.3). In total, 50 per cent of the construction costs were supplied in kind.[134] The process to agree on in-kind contributions and to manage delivery and implementation was complex. It included a proposal for in-kind contributions that needed to be reviewed by the European XFEL IKRC (In-Kind Review Committee) and approved by the management board or council, depending on the value of the contribution. The eventual design, manufacturing, delivery and implementation, moreover, required an elaborated time schedule, and clear communication and definition of work packages.[135] France, for instance, only contributes in kind. Russia, on the contrary, only contributes in cash, but Russian institutes develop and construct components and equipment for the project that are based on manufacturing contracts.[136] For the contributing partner, in-kind contributions are a way to develop and improve its own expertise and knowledge as well as to boost the national economy. Goods are, for instance, manufactured by the contributing partner's national companies and institutes, and staff is sent abroad to work at the project for a couple of years to improve knowledge, and for skills and expertise to be brought back to the home country. Certainly, to the extent that in-kind contributions also signify a delegation of responsibilities and risks from the project to the contributing partner, this can also cause failure to meet the expected standards, as well as overall delay for the project.[137]

5.6 The Role of Russia

Russia became the second biggest shareholder in the European XFEL project contributing 250 million euros, corresponding to 23.1 per cent of the construction costs. Russia also became a member in the FAIR project with a similar substantial contribution of 178 million euros, corresponding to 17.3 per cent of the construction costs.[138] Taken together, Russian contributions to these two facilities make up "the first time that Russia is investing such amounts of financing in research facilities which are not located on its own territory."[139] It can be argued that such large investments are symptomatic for a wider and thorough re-orientation of Russian science policy and the re-structuring of its science system following the economic and political upheavals in the 1990s.

The Soviet Union had developed a large science system and a high-level scientific infrastructure. However, after the dissolution of the Soviet Union in the early 1990s, investments in economic and political stability were prioritised while national spending on science and technology in Russia decreased dramatically. Russian domestic spending on science and technology (in per cent of GDP) fell from 1.91 per cent in 1990 to 0.69 per cent in 1992 and 0.72 per cent in 1993. Between 1990 and 2009, spending on science and technology (in per cent of GDP) in France, Germany and the United States oscillated in each of these countries between 2 and 3.5 per cent. The United Kingdom spent between 1.5 and 2 per cent of the GDP in the same time range.[140] This overall lack of funding and political support, together with ageing researchers due to a massive emigration of young and skilled scientists and outdated equipment in most research institutes, called the future of Russian science at that time seriously into question.[141]

It was not only before president Vladimir Putin's second term in office (2004–2008) that Russia's visible strive towards increasing political and economic influence on the international agenda also translated to the fields of science, technology and research. On the one hand, several research policy tools that were concluded during that time were expected to lift the country on competitive footing with regard to its political and economic competitors in a globalised and increasingly knowledge-based world.[142] These included, for instance, large investments into domestic science and technology initiatives, such as the implementation of Federal Target Programmes with a main topical focus on nanotechnology, or the creation of several joint stock companies such as ROSATOM (Rosatom State Nuclear Energy Corporation) or RUSNANO (Russian Corporation of Nanotechnologies) that manage national research and development (R&D) projects and Russian membership in international collaborations.[143] On the other hand, the accession of Russia as a new member to several collaborative Big Science projects in Europe since the late 1990s, which was accompanied by substantial financial contributions and long-term commitment, adds another aspect. Russia's growing involvement in (Western) European Big Science can certainly be interpreted as an additional way to strengthen expertise and capabilities of the country in cutting-edge research. But there is also much to suggest that it was a way to consolidate and project political power in Europe. In other words, the large Russian contribution to the European XFEL project was not only a means to get access to the state-of-the-art research facilities and experimental equipment but also to create long-term and mutually dependent relations to European countries. It can be assumed that these relations not only challenged periods of diplomatic frost and political disarray but also strengthened the German-Russian bilateral relations that can be regarded as an important political cornerstone on the European continent.

5.6.1 *German-Russian Collaborations in Science*

The end of the Cold War, the fall of the Berlin Wall and the dissolution of the Soviet Union in the early 1990s had swept away political certainties and geopolitical balances. In as much as Germany and Russia found themselves in a vital period of change and transition beginning in the 1990s, both countries also looked for ways to define common ground for cooperation and stable diplomatic relations. The heads of governments, from chancellor Helmut Kohl and president Boris Yeltsin, to chancellor Gerhard Schröder and president Vladimir Putin, frequently demonstrated amicable closeness. German-Russian bilateral relations seemed to thrive during the 1990s and early 2000s.[144] At the same time, Russian foreign policy concepts and diplomacy often operated on two different levels, between intentions and expectations and between rhetoric and reality. With "Russia defined as a great power in one paragraph and as fundamentally pragmatic in the next,"[145] this two-track approach often led to diverging and confusing signals within bi- and multilateral contexts.[146]

Several bilateral agreements that aimed at deepening German-Russian relations were concluded over the course of the 1990s. Initiated by Schröder and Putin, the heads of government meet for the *German-Russian Intergovernmental Consultations* once every year since 1998. The *Joint Declaration on Cooperation in Modernisation Partnership* and the *Petersburg Dialogue* were initiated in 2001. The *Common Agreement on Strategic Partnership in Education, Research, and Innovation* that was signed in April 2005 demonstrated the political willingness to further strengthen the bilateral relations in science and research cooperation.[147] Interdepartmental agreements between the German BMBF and the Russian Ministry for Education and Science amended these bilateral agreements, and covered a broad variety of thematic fields, such as laser research, marine and polar research, biological research and accelerator-based photon sources.[148]

Related to this context, new structures, supporting schemes and instruments were set up both within the bilateral German-Russian and the multilateral EU-Russian context to foster and strengthen cooperation, and to facilitate the mobility of researchers, co-publications and the creation of joint committees and working groups.[149] Russia was the most successful non-EU country in the seventh FP of the EU (2007–2013), and there is much to suggest that such a strong relationship translated into the 2014 EU-Russian Year of Science.[150] As a visible testament to increasing bilateral scientific cooperation, the DFG (Deutsche Forschungsgemeinschaft, German Research Foundation) and the Helmholtz Association established offices in Moscow in 2003 and 2005, respectively. The German Historical Institute opened in Moscow in 2005. The year 2011/12 became the German-Russian Year of Science.[151]

The 2005 *Common Agreement on Strategic Partnership in Education, Research, and Innovation*[152] as well as the major scientific and technical contributions of Russian scientists to the development of free-electron laser technology constitute the two major contexts in which the Russian participation in the European XFEL needs to be understood. Interestingly, in 1991 and 1992, two interdepartmental agreements between the German BMBF and the Russian Ministry of Education and Science on high-temperature superconductivity as well as laser research and laser technology were concluded that refrain crucial technological concerns around both the TESLA linear collider and the free-electron laser proposal.[153]

The 2005 agreement can be characterised as an ad-hoc and top-down decision by the two heads of government, without preparatory activities from the German or Russian governmental bodies and ministries. There is much to suggest that it was merely a symbolic statement of intention that lacked concrete measures and operational clout. To fill this agreement with life, then state secretary Frieder Meyer-Krahmer at the German BMBF proposed to invite Russia as a member to the European XFEL project. In 2005, an informal meeting took place in Moscow between stakeholders of the BMBF and the Russian Minister of Education and Science Andrei Fursenko and Mikhail Kovalchuk, who was at that time scientific secretary of the Russian Council for Science and High Technologies, a governing body directly linked and subordinated to the Russian president.[154] Soon after, Russia signed the European XFEL Memorandum of Understanding. Meyer-Krahmer's initiative can, on the one hand, be interpreted as a way to underpin the rhetoric of the 2005 agreement with a binding commitment that would tie the research policies and Big Science strategies of the two countries together for several decades to come. On the other hand, it can also be regarded as a means to acquire an additional foreign contribution to the European XFEL project, after the commitment of other European countries fell short of German expectations.[155]

Shortly after the signing of the 2005 agreement between Gerhard Schröder and Vladimir Putin, the German government changed from a socialist-green coalition to the grand coalition of social democrats and conservatives. Competences in the BMBF changed from socialist minister Edelgard Bulmahn to conservative minister Annette Schavan. Work on the agreement was nevertheless continued.[156] This is worth mentioning because the biography of the new conservative chancellor Angela Merkel clearly did not facilitate the creation of a strong sense of personal closeness with president Vladimir Putin, as it had been in the case of chancellor

Gerhard Schröder. Born in Hamburg, West Germany, but raised in East Germany as the daughter of a protestant priest, Merkel resisted Soviet Union communism in political opposition.[157] She adopted a value-based foreign policy approach towards Russia that would preside over (economic) interests, opposing the approach of her predecessor in office Gerhard Schröder.

The official agreement on Russian membership in the European XFEL project was concluded on a high political level between Merkel and Putin during the German-Russian Government Consultations at Wiesbaden in October 2007. The 2007 summit did not continue the seemingly amicable relationship between Schröder and Putin. On the contrary, the meeting between Merkel and Putin was described as a "cold encounter."[158] In addition, the bilateral agreement on the development and application of accelerator-based photon sources, signed in 2007, coincided with and supplemented the Russian commitment to the European XFEL project. In 2012, the two countries also established IRI (Ioffe-Röntgen Institute) as an additional effort to enhance bilateral cooperation in Big Science (see also Fig. 5.2).[159] Taken together, these efforts clearly mirror the strategic significance of collaboration in photon science for the German-Russian relations in science and technology in the early twenty-first century.

5.6.2 *Nanotechnology, Big Politics and the European XFEL*

To the extent that research at the free-electron laser of the European XFEL was also expected to improve insights in nanostructures and to advance application of nanomaterials and nanotechnologies, the decision of the Russian government to participate in the European XFEL project also needs to be related to increasing national efforts and investments in nanotechnology programmes. On the one hand, these measures since the mid-2000s were certainly triggered by developments in other countries, such as the United States and China. As these countries were considered Russia's main economic and political competitors in the early twenty-first century, science and technology efforts and developments within these two countries always carried a strong strategic rationale.[160] On the other hand, and related to the first aspect, there has been a veritable political hype around nanotechnologies and research on nanomaterials and nanostructures in recent decades. This way mainly due to promises and rhetoric that signalled enormous impacts that this innovative technology may have

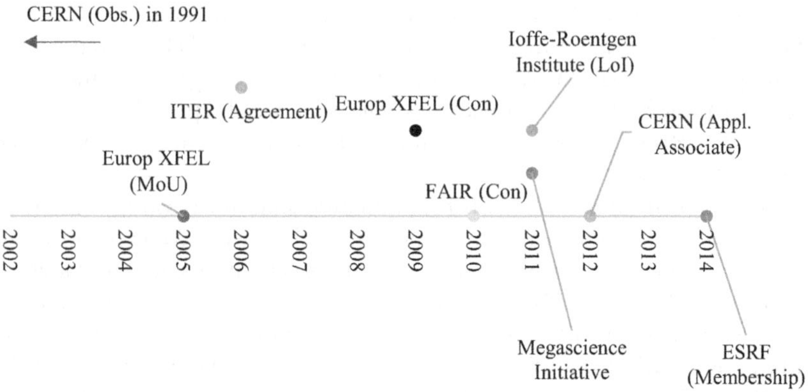

Fig. 5.2 Russian involvement in Big Science in Europe, 1991–2014. (Note: Con: Convention, MoU: Memorandum of Understanding, LoI: Letter of Intent, Appl.: Application, Obs.: Observer). (Sources: Letter of Intent of the Deutsches Elektronen-Synchrotron DESY (Hamburg, Germany) and the National Research Centre "Kurchatov Institute" (Moscow, Russian Federation) Concerning the German-Russian Cooperation in Strategic Research Fields: The Ioffe-Röntgen Institute (May 23, 2011); Protocole d'Adhésion du Gouvernment de la Féderation de Russie à la Convention du 16 Décembre 1988 Relative à la Construction et à l'Exploitation d'une Installation Européenee de Rayonnement de Synchrotron (June 23, 2014); Convention concerning the Construction and Operation of a Facility for Antiproton and Ion Research in Europe (FAIR) (October 12, 2010); Convention concerning the Construction and Operation of a European X-Ray Free-Electron Laser Facility; Agreement on the Establishment of the ITER International Fusion Energy Organization for the Joint Implementation of the ITER Project (November 21, 2006); S. Smirnov, Russia – CERN Cooperation: Current Status and Perspectives, *Journal of Physics: Conference Series*, 1406, no. 1 (2019))

on every aspect of life in the near future, ranging from the manipulation to re-invention of surfaces, drugs or molecules.[161]

Political stakes were high for the approval and implementation of several government-sponsored initiatives on nanoscience and nanotechnology in Russia. For instance, a Federal Target Program on *Development of the Nano-Industry Infrastructure in the Russian Federation* as well as a development programme for the *Nano-Industry in the Russian Federation to the year 2015* were implemented in 2007. In February

2008, the Russian Minister for Science and Education Andrei Fursenko declared as follows: "We are practically ready for a new scientific and technological revolution: a nanotechnology revolution."[162] Investments into this emerging and highly promising field were expected to reduce the dependency of the country on oil and gas, the prices of which had shortly before started to drop in the dawn of a deepening global economic crisis. Nanotechnology and nanoscience were also regarded as a promising and financially rewarding way to strengthen Russian capacities and capabilities in commercialisation, production and application, which remained weak by the mid-2000s despite growing financial support.[163] The creation of RUSNANO (Russian Corporation of Nanotechnologies) in 2007 was also related to these developments. RUSNANO manages and coordinates the national activities and initiatives in nanotechnology. Until 2016, it also oversaw the Russian contributions to the European XFEL project. RUSNANO, similar to ROSATOM, also acts as a governmental agent, and the Russian government exercises control over the appointment of leading positions. ROSATOM runs the Russian nuclear facilities and manages, among others, the Russian contributions to the FAIR and ITER projects. RUSNANO also supports application-oriented nanotechnology projects and assesses the potential for commercialisation.[164] Successful project applications were required to have production and manufacturing facilities in Russia. In contrast to large nanotechnology and nanoscience programmes in other countries, the Russian nanotechnology initiatives were thus not primarily based on international outreach and brain-gain from abroad but on the comprehensive support of existing domestic structures and locations in Russia.[165]

Russian participation in the European XFEL project in the context of its national nanotechnology initiatives did not lack political clout. Mikhail Kovalchuk and Andrei Fursenko, who met the delegation from the German BMBF in 2005, also played key roles in preparing and implementing nanotechnology initiatives. Kovalchuk appeared to have convinced President Putin to create a large programme in nanotechnology and to provide generous funding. Kovalchuk and Fursenko both acquired high professional positions within the 2007 nanotechnology programme: Fursenko as deputy supervisor of the programme and Kovalchuk as director of the Kurchatov Institute, which became charged with the scientific coordination of the nanotechnology programme in 2007.[166]

Another strategic aspect of Russia's involvement in the European XFEL project is the country's increasing participation in collaborative Big Science projects in Europe since the late 1990s. In addition to FAIR and the European XFEL project, Russia is a full member of the ITER project. The history of ITER dates to the Cold War fragile détente of the 1970s when the project idea was launched to initiate scientific cooperation between the United States and the Soviet Union.[167] In 1999, Russia acquired observer status at CERN and in 2012, the country applied for associate membership. The 2012 application was, however, also enforced by the fact that in 2010, CERN had decided to outrun the observer model but to offer its countries the possibility to become associate or full members.[168] In 2014, Russia became a full member of the ESRF, contributing 6 per cent of the operation costs, which led to a large re-allocation process by other member countries of the ESRF.

The Russian Megascience Initiative, which was set up in 2011, is related to these developments (see also Table 5.2). This initiative aims to establish several Big Science facilities with international contributions on Russian

Table 5.2 Projects within the Russian Megascience Initiative and corresponding facilities in Europe

Facilities proposed by Russia (planned or under construction on Russian territory)	*Facilities in (Western) Europe (planned, under construction or in operation)*
Scientific and Research Reactor Complex (PIK)	Institut Laue-Langevin (ILL), Grenoble (France)
Nuclotron-Based Ion Collider Facility (NICA)	Facility for Antiproton and Ion Research (FAIR) at GSI, Darmstadt (Germany)
Fourth Generation Special-Purpose Synchrotron Radiation Source (SSRS-4)	PETRA IV at DESY, Hamburg (Germany)
Exawatt Center for Extreme Light Studies (XCELS)	Extreme Light Infrastructure (ELI), Czech Republic, Romania and Hungary; European XFEL, Hamburg (Germany)
Tokamak Fusion Reactor (IGNITOR)	International Thermonuclear Experimental Reactor ITER, Cadarache, (France)
Super Tau-Charm Factory (STC)	CERN, Geneva (Switzerland)

Source: Website of the CREMLIN Project (Connecting Russian and European Measures for Large-Scale Research Infrastructures): https://www.cremlin.eu

territory. But, importantly, the projects included in this initiative mirror those in Western Europe scientifically and technologically (see Table 5.2).[169] With regard to Russia's large contributions to European Big Science projects in recent decades, this initiative certainly was a means for Russia to claim a return on investment from European countries. The EU-funded and DESY-coordinated CREMLIN (Connecting Russian and European Measures for Large-Scale Research Infrastructures) project is telling in this regard. It aims to advance cooperation between the Russian projects and its European counterparts and to transform this seemingly doubling effort (scientifically and financially) into cooperative complementary action.[170] Although the Megascience Initiative became effective only in 2011, there is much to suggest that preparation activities were clearly rooted in the atmosphere of political change in the mid-2000s.

Taken together, these manifold developments in Russia such as increasing investments in nanotechnology, the implementation of the Megascience Initiative as well as growing participation in Western European Big Science projects (including the large financial contribution to the European XFEL) can be interpreted as a harbinger of future Russian activities in science policy and high politics that certainly will profoundly impact and shape Big Science and big politics in (Western) Europe in the coming years and decades.

5.7 Towards a Convention

In June 2007, the thirteen signatories of the Memorandum of Understanding of 2004/2005 signed a *Communiqué on the Official Launch of the European XFEL*. In the same year in June, the EU-funded Pre-XFEL project also started under the auspices of the seventh FP of the EU. With a duration of three years and a coordination in the hands of DESY, its core aim was to provide overall support for the foundation of the European XFEL, meaning the preparation of the legal and organisational framework, as well as the recruitment of staff and the attraction of potential users.[171] However, the involvement of the EU and/or European Commission in the European XFEL project is very limited. The project receives funding through the Pre-XFEL project under the auspices of the seventh FP of the EU. It is moreover represented in several European networks and initiatives that were partly funded through the FP of the EU, or that maintained close connections with the European Commission.[172]

Final negotiations on the financial shares among the different national delegations resulted in a very unequal distribution of the construction costs. The large contribution made Russia a heavyweight on the negotiation table that resulted in changes in the organisational structure of the European XFEL project and the voting system, as well as the way the facility is used and accessed. Concerning the organisation of the facility, the position of a distinguished advisor to the Administrative Director of the European XFEL GmbH was created in November 2010. Since then the position is staffed by Mikhail Rychev, a former employee of the Russian Kurchatov Institute. This additional position undoubtedly strengthened the Russian communication with the management board of the European XFEL, where only one (scientific director) out of five positions (two managing directors and three scientific directors) was and remains filled by a person of Russian nationality.[173]

The voting procedure at the European XFEL is based on a mixed system that considers the share capital of the member country, as well as the total number of the contracting parties. Different quotas were defined for simple and qualified majority voting as well as for unanimous decision. The case for the qualified majority voting is particularly interesting, as it equals "at least 77% of the share capital and the Shareholders of not more than half of the Contracting Parties voting against."[174] Germany and Russia together hold 76.71 per cent of the shares. In this regard, on the one hand, the two countries need at least a third country to take decisions based on a qualified majority. One the other hand, it also seems as if Russia ensured to hold a share that is large enough that no qualified majority decision can be taken without it. Regarding the 77 per cent of the share capital needed to make a qualified majority decision, Russia would need a share of at least 23.1 per cent to get a de-facto veto right in qualified majority voting under the condition that no more than half of the contracting parties vote against them. And indeed, Russia's contribution of about 250 million euros equals exactly the required 23.1 per cent of the total shares. The *Russian declaration on the Final Act of the Conference of Plenipotentiaries for the Establishment of a European X-Ray Free-Electron Laser Facility* further supports this line of argumentation:

> The share of the Russian Shareholder in the capital of the Company must ensure a volume of voting rights whereby, without approval from the shareholder, no decision which requires a qualified majority according to the Articles of the Association of the Company may be taken by the Council of the Company.[175]

The same declaration also reads that "the list of matters subject to approval by a qualified majority shall in any case remain unchanged."[176] Through its on-the-point contribution of 23.1 per cent, Russia seemingly has established a crucial mechanism: It is in full control both of the issues that are subject to qualified majority voting and the decision-making process.

A similar demonstration of Russia's powerful role concerns the use of, and access to, the facility. Article 6 of the European XFEL convention mentions that "[t]he use of the European XFEL Facility shall be based on criteria of scientific excellence and benefits to society." This is a common modus operandi for publicly funded intergovernmental Big Science projects: User groups apply for experimental time based on a scientific peer-review process.[177] Although this constitutes a well-established mechanism, it remains an ideal that has been subject to conflict-filled situations among the participating members of the facilities. Imbalances occurred when the scientific communities of the participating countries got too little or too much experimental time at the beam lines compared to the corresponding share of their country, that is, when scientists over-use or under-use the experimental resources compared to the corresponding shares of their home countries in the overall project.[178] In order to address the potential tensions that arise from this well-known phenomenon, the European XFEL convention states that the:

> [c]ouncil creates the prerequisites to avoid a lasting and significant imbalance between the use made of the European XFEL Facility by the scientific community of a Contracting Party country and the contribution of that Party's Shareholder(s) to the European XFEL Facility.[179]

Moreover, this provides the background against which an additional Russian statement on processes and rules foreseen in the convention needs to be interpreted:

> [T]o stipulate the level of the Russian Federation's share in the operating costs (...), it must be borne in mind that in line with the principle of proportionality this is to be calculated on the basis of the period of time the facility is used by scientists of Russian research organizations.[180]

This statement can be read in two ways. First, Russia acknowledges the possibility and the need to adjust contributions when scientists regularly or over a long time over-use or under-use the facility compared to the contributions of their governments, which would stand in line with the above-mentioned provisions made in the European XFEL convention. A second possibility is, however, that access to the facility for Russian scientists should be based, first and foremost, on the Russian financial shares in this project. With regard to additional funding that Russia is willing (and seemingly able) to spend on this project, their scientists may get greater parts of experimental time than granted by the scientific review process. A press release from RUSNANO in December 2009, several days before the European XFEL convention was signed, further supports this perspective. It stated that "beam usage time will be shared proportionally to each country's contribution to the project."[181]

Furthermore, it needs to be highlighted that organisational parallels between HERA and the European XFEL should not be underestimated.[182] Similar to HERA, foreign partners contributed to the project through in-kind contributions. While in-kind contributions to the accelerator complex were a novelty for the particle physics community when the HERA project was proposed, over the last years, in-kind contributions became an attractive way for countries to contribute to international collaborations. Certainly, the European XFEL, by its convention in 2009, became an independent and self-standing collaborative facility under German law based on intergovernmental agreement. Its management and organisation included negotiation and agreement on several matters of international law, such as VAT (Value Added Tax), the status of employees or international schooling. Based on intergovernmental agreement, the European XFEL project is no longer one of the projects of DESY as it has been the case for HERA. But the local anchoring of the European XFEL at Hamburg and Schenefeld, the major German contribution and the decisive role of DESY in managing various aspects of the preparatory activities and its role in construction constituted a situation similar to that of HERA several decades ago. In other words, the scientific and technological promises of the European XFEL project needed to be attractive enough to make foreign countries contribute to a facility that would be dominated by German political commitment and financial contribution and managed in (many regards) by DESY and that never seriously considered to search for another site than Hamburg/Schenefeld.[183]

5.8 Concluding Discussion

The European XFEL project originated as a side-branch of the international TESLA collaboration at DESY, which had initially proposed a linear collider in particle physics in the early 1990s. However, the 2003 decision of the German government changed the plans of the TESLA collaboration because it put a halt to the linear collider effort but supported the realisation of the free-electron laser. On the one hand, this was an unexpected drawback for the TESLA collaboration and the research centre DESY that considered the linear collider as the core concern of the project proposal and the free-electron laser as a side-branch only. On the other hand, the go-ahead of the Germany government for the free-electron laser was a major aspect, but only the most recent one, in the gradual transformation of DESY from a single-mission particle physics research centre to a multidisciplinary facility with the main focus on photon science.

The early history of the European XFEL built entirely on DESY and the TESLA collaboration. The development and transformation of DESY over the course of the last decades, and its overall successful scientific standing in national, European and international contexts was one fundamental building block of Germany's large involvement in the European XFEL project. Until today, DESY remains closely connected to this project through a specific long-term agreement, which stipulates that DESY is not only the host for the injector complex of the European XFEL but also the key actor in the construction and operation of the free-electron laser's accelerator complex.[184]

Financially, the European XFEL project became dominated by Germany and Russia, who together contribute 76.1 per cent of the construction costs. Financial dominance does not resonate a genuine bilateral partnership, but the individual political rationales behind the large contributions from both countries can hardly be brought into conjunction. For Germany as well as for Russia, divergent national strategies triumphed over common and concerted efforts, and the case of the European XFEL illustrates that in science, as much as in post-Cold War politics, the two partners sit uneasily together.

For Germany, the 2003 decision to realise the X-ray free-electron laser and to put a halt to the development of the linear collider was an important strategic move within the broader national science policy context, resonating a re-framed priority-setting in Big Science and research towards multidisciplinary, application-oriented research on Big Science user facilities. For Russia, the large contribution to the European XFEL was an additional step in its increasing involvement in Big Science in Europe since the late 1990s. This development was (seen retrospectively) only a harbinger of the Russian 2011 Megascience Initiative, by which the country required and urged European countries a return on investment to large-scale research facilities that are scientifically and technologically mirroring those in Europe, but that should be built on Russian territory in the near future.

Notes

1. See Art 1.3 of the *Convention concerning the Construction and Operation of a European X-Ray Free-Electron Laser Facility,* November 30, 2009; Art. 2 of the *Articles of Association of the "European X-Ray Free-Electron Laser Facility GmbH" (European XFEL GmbH),* November 30, 2009.
2. See, for example, T. Heinze, O. Hallonsten, and S. Heinecke, "From Periphery to Center: Synchrotron Radiation at DESY, Part I: 1962–1977." *Historical Studies in the Natural Sciences* 45, no. 3 (2015); T. Heinze, O. Hallonsten, and S. Heinecke, "From Periphery to Center: Synchrotron Radiation at DESY, Part II: 1977–1993." *Historical Studies in the Natural Sciences* 45, no. 4 (2015); T. Heinze, O. Hallonsten, and S. Heinecke, "Turning the Ship: The Transformation of DESY, 1993–2009." *Physics in Perspective* 19, no. 4 (2017); E. Lohrmann and P. Söding, *Von schnellen Teilchen und hellem Licht: 50 Jahre Deutsches Elektronen-Synchrotron DESY* (Weinheim: Wiley, 2009).
3. See, for example, Heinze, Hallonsten, and Heinecke, "From Periphery to Center I"; Heinze, Hallonsten, and Heinecke, "From Periphery to Center II"; Heinze, Hallonsten, and Heinecke "Turning the Ship."
4. See, for example, L. Hoddeson, L. M. Brown, M. Dresden, and M. Riordan (eds.), *The Rise of the Standard Model: A History of Particle Physics from 1964 to 1979* (Cambridge: Cambridge University Press, 1997); P. Galison and B. Hevly, *Big Science. The Growth of Large-Scale Research* (Stanford: Stanford University Press, 1992).
5. See, for example, M. Lengwiler, "Kontinuitäten und Umbrüche in der Deutschen Wissenschaftspolitik des 20. Jahrhunderts." In *Handbuch Wissenschaftspolitik,* eds. S. Hornbostel, A. Knie, and D. Simon (Wiesbaden: Verlag für Sozialwissenschaften, 2010); G. Ritter, *Großforschung und Staat in Deutschland: Ein historischer Überblick* (München: Beck, 1992); J. Krige, *American Hegemony and the Postwar Reconstruction of Science in Europe* (Cambridge, MA.: MIT Press, 2006).

6. See, for example, C. Habfast, *Grossforschung mit kleinen Teilchen: Das Deutsche Elektronen-Synchrotron, DESY, 1956–1970* (Berlin, New York: Springer, 1989); Lohrmann and Söding, *Von schnellen Teilchen*; Heinze, Hallonsten, and Heinecke, "From Periphery to Center I", 458–459.
7. See Lohrmann and Söding, *Von schnellen Teilchen*, 17–18.
8. See Lohrmann and Söding, *Von schnellen Teilchen*, 4; E. Lohrmann and P. Söding, "DESY Marks 50 Years of Accelerator Research." *CERN Courier*, December 7, 2009.
9. European Committee for Future Accelerators, *Report 1967* (Geneva: CERN, 1967), 31, Table X.
10. Heinze, Hallonsten, and Heinecke, "From Periphery to Center I."
11. Lohrmann and Söding, *Von schnellen Teilchen*, 119–122.
12. Lohrmann and Söding, *Von schnellen Teilchen*, 116–119; Heinze, Hallonsten, and Heinecke, "From Periphery to Center I."
13. Lohrmann and Söding, *Von schnellen Teilchen*, 126–132, 138–141; C. Harringa, "Zwischen Völkerrecht und Frascati: Praktische Aspekte der Ausgestaltung internationaler Kooperationen am Deutschen Elektronen Synchrotron (DESY)." *Ordnung der Wissenschaft* 2 (2018).
14. See, for example, R. Siemann, "Uses of Superconductivity in Particle Accelerators." In *Lepton – Hadron Scattering, Proceedings*, 19th SLAC Summer Institute on Particle Physics (SSI 91), ed. J. Hawthorne (Stanford: SLAC).
15. See *Bericht des Direktoriums zur 102. Sitzung des DESY Verwaltungsrates am 2./3. Dezember 2004 bei DESY in Hamburg*, DESY Archive; T. Behnke et al., eds., *The International Linear Collider: Technical Design Report*, 2013.
16. See, for example, Heinze, Hallonsten, and Heinecke, "From Periphery to Center I"; Lohrmann and Söding, *Von schnellen Teilchen*, 221–227; C. Kunz, *Synchrotronstrahlung bei DESY: Anfänge* (private print, 2012).
17. See Heinze, Hallonsten, and Heinecke, "From Periphery to Center I", 472 f.
18. See Heinze, Hallonsten, and Heinecke, "From Periphery to Center I", 476–477.
19. See Heinze, Hallonsten, and Heinecke, "From Periphery to Center II", 581ff.
20. See Heinze, Hallonsten, and Heinecke, "From Periphery to Center II". Appendix/Table 3.
21. See, for example, Heinze, Hallonsten, and Heinecke, "Turning the Ship"; Lohrmann and Söding, *Von schnellen Teilchen*, 221–254; *Niederschrift über die 95. Sitzung des Verwaltungsrates am 5.12.2001 in Hamburg*, DESY Archive.

22. See DESY, *Particle Physics 2016: Jahresbericht des Bereichs Teilchenphysik bei DESY* (Hamburg: DESY, 2017).
23. See, for example, *100. Sitzung des DESY Verwaltungsrates am 24./25. Juni 2004 bei DESY in Zeuthen*, DESY Archive; Heinze, Hallonsten, and Heinecke, "Turning the Ship", 440.
24. TESLA Collaboration, *A Proposal to Construct and Test Prototype Superconducting RF Structures for Linear Colliders* (Hamburg: DESY, 1992).
25. See Lohrmann and Söding, *Von schnellen Teilchen*, 299–300.
26. See R. Brinkmann et al., *Conceptual Design Report of a 500 GeV $e^+ e^-$ Linear Collider with Integrated X-Ray Laser Facility* (Hamburg: DESY, 1997), 281; Heinze, Hallonsten, and Heinecke, "Turning the Ship."
27. For pioneering efforts at the US-American Fermilab in the 1970s of superconducting technology in a circular-shaped accelerator, see L. Hoddeson, "The First Large-Scale Application of Superconductivity: The Fermilab Energy Doubler, 1972–1983." *Historical Studies in the Physical and Biological Sciences* 18, no. 1 (1987).
28. See Lohrmann and Söding, *Von schnellen Teilchen*, 299–300; Brinkmann et al., *Conceptual Design Report*, 597–598.
29. See *Niederschrift über die 87. Sitzung des Verwaltungsrates am 12. November 1997 in Zeuthen*, DESY Archive; DESY, *Wissenschaftlicher Jahresbericht 1998* (Hamburg: DESY, 1999), 9.
30. See *DESY, Planungsschritte Linearcollider*, DESY Archive.
31. See *DESY, Planungsschritte Linearcollider*, DESY Archive.
32. See Bundestag, *Antrag der Abgeordneten Ulrike Flach* et al. *und der Fraktion der F.D.P.: Forschungsstandort Deutschland stärken – Zukunftsprojekt "TESLA" nicht gefährden*, Drucksache 14/4646, November 15, 2000.
33. See DESY, *Planungsschritte Linearcollider*, DESY Archive.
34. See DESY, *Wissenschaftlicher Jahresbericht 1998*.
35. See Bürgerschaft der Freien und Hansestadt Hamburg, *Mitteilung des Senats an die Bürgerschaft. Gesetz zum Staatsvertrag zwischen der Freien und Hansestadt Hamburg und dem Land Schleswig-Holstein über die Schaffung der planerischen Voraussetzungen für die Errichtung und den Betrieb eines Linearbeschleunigers,* Drucksache 16/436, November 4, 1998.
36. See Bürgerschaft der Freien und Hansestadt Hamburg, *Gesetz zum Staatsvertrag*, November 4, 1998.
37. See, for example, R. Brinkmann, P. Schmüser, and B. Wiik, "TESLA." *Physik in unserer Zeit* 30, no. 2 (1999), 62; Wissenschaftsrat, *Stellungnahme zu neun Großgeräten der naturwissenschaftlichen Grundlagenforschung und zur Weiterentwicklung der Investitionsplanung von Großgeräten /*

Statement on Nine Large-Scale Facilities for Basic Scientific Research and on the Development of Investment Planning for Large-Scale Facilities (Köln: Wissenschaftsrat, 2003), 220.

38. See Wissenschaftsrat, *Stellungnahme zu neun Großgeräten*, 220.
39. See Wissenschaftsrat, *Stellungnahme zu neun Großgeräten*, 220.
40. See Wissenschaftsrat, *Stellungnahme zu neun Großgeräten*, 220.
41. See ICFA, *Statement by the International Committee for Future Accelerators 6 May 1993 and Interlaboratory Memorandum of Understanding of Research and Development Towards TeV Linear e^+e^- Colliders by DESY, KEK and SLAC*, annexed to: *TESLA Collaboration, TTF Collaboration Meeting-Final Report*, June 1–2, 1993, Hamburg; ICFA Statement from May 6, 1993: "Research and Development Towards TeV Linear e^+e^- Colliders." In *Niederschrift über die 78. Sitzung des Verwaltugsrates am 13. Juli 1993 in Zeuthen*, DESY Archive.
42. See Riordan, Hoddeson, and Kolb, *Tunnel Visions*, 26, 258 fn. 52.
43. See Riordan, Hoddeson, and Kolb, *Tunnel Visions*, ix.
44. See L. Maiani and L. Bonolis, "The LHC Timeline: A Personal Recollection (1980–2012)." *The European Physical Journal* 42, no. 4–5 (2017), 482.
45. See, for example, D. Kevles, "Good-Bye to the SSC: On the Life and Death of the Superconducting Super Collider." *Engineering & Science* 58, no. 2 (1995); M. Riordan, "A Bridge Too Far: The Demise of the Superconducting Super Collider." *Physics Today* 69, no. 10 (2016).
46. O. Hallonsten, *Big Science Transformed: Science, Politics and Organization in Europe and the United States* (Cham: Palgrave Macmillan, 2016), 32.
47. See Riordan, Hoddeson, and Kolb, *Tunnel Visions*, 256.
48. See Riordan, Hoddeson, and Kolb, *Tunnel Visions*, 256; Hallonsten, *Big Science Transformed*, 66–67.
49. See H. Schunck, *Meine Jahre mit DESY* (private print, 2014), 8.
50. See Heinze, Hallonsten, and Heinecke, "Turning the Ship", 433.
51. J. Schneider, "FLASH – From Accelerator Test Facility to the First Single-Pass Soft X-Ray Free-Electron Laser," *Journal of Physics B* 43, no. 19 (2010), 2.
52. See, for example, D. Greenberg, *The Politics of Pure Science* (Chicago: University of Chicago Press, 1999); Hallonsten, *Big Science Transformed;* N. Oreskes and J. Krige (eds.), *Science and Technology in the Global Cold War* (Cambridge, MA: MIT Press, 2014); Krige, *American Hegemony.*
53. See, for example, S. Shapin, *The Scientific Life: A Moral History of a Late Modern Vocation* (Chicago: University of Chicago Press, 2008); N. Stehr, *Knowledge Societies* (London: SAGE, 1994); M. Castells and G. Cardoso (eds.) *The Network Society: From Knowledge to Policy* (Washington DC: Johns Hopkins Center for Transatlantic Relations, 2005); E. Kallerud

et al., *Dimensions of Research and Innovation Policies to Address Grand and Global Challenges* (NIFU Working Paper 13, 2013); Hallonsten, *Big Science Transformed.*

54. Hallonsten, *Big Science Transformed*, 3.
55. Hallonsten, *Big Science Transformed*, 213.
56. See Hallonsten, *Big Science Transformed*, 10.
57. See Wissenschaftsrat, *Perspektiven für Wissenschaft und Forschung auf dem Weg zur deutschen Einheit: Zwölf Empfehlungen* (Berlin: Wissenschaftsrat, 1990).
58. See O. Hallonsten and T. Heinze, "Institutional Persistence Through Gradual Organizational Adaptation: Analysis of National Laboratories in the USA and Germany," *Science and Public Policy* 39, no. 4 (2012), 456.
59. See Heinze, Hallonsten, and Heinecke, "Turning the Ship", 429 (Figure 1) and 438.
60. Heinze, Hallonsten, and Heinecke, "Turning the Ship", 435.
61. There are mainly two kinds of linac-based light sources with the characteristics of a laser: FEL (free-electron laser) and ERL (energy recovery linac). Only the FEL will be discussed here.
62. See, for example, H. Winick, *Synchrotron Radiation Sources: Present Capabilities and Future Directions* (SLAC-PUB-777 1), 1998, 1; H.-D. Nuhn, "From Storage Rings to Free Electron Lasers for Hard X-Rays," *Journal of Physics: Condensed Matter* 16, no. 33 (2004), 16.
63. See, for example, FLASH at DESY, FELIX (Free Electron Lasers for Infrared Experiments) in Nijmegen, the Netherlands, FELBE (Free-Electron Laser at the Electron Linear Accelerator with High Brilliance and Low Emittance) at the Helmholtz-Zentrum Dresden-Rossendorf, Germany, POLFEL (Polish Free Electron Laser), MAX IV FEL near Lund, Sweden as part of the MAX IV laboratory, FERMI (Free Electron Laser Radiation for Multidisciplinary Investigations) at the Elettra Synchrotron in Trieste, Italy, and CLIO (Centre Laser Infrarouge d'Orsay) near Paris, France.
64. See DESY, *Memorandum on the Realisation of the First Stage of the European TESLA XFEL LABORATORY, 24 October 2002*, DESY Archive.
65. See *Niederschrift über die 82. Sitzung des Verwaltungsrates am 28.6.1995 in Zeuthen*, DESY Archive.
66. See C. Pellegrini, "The History of X-Ray Free-Electron Lasers," *The European Physical Journal* 37, no. 5 (2012).
67. Heinze, Hallonsten, and Heinecke, "Turning the Ship", 433.
68. See *DESY, Planungsschritte Linearcollider*, DESY Archive; DESY, *Wissenschaftlicher Jahresbericht 1995* (Hamburg: DESY, 1996), 225–234.

69. See DESY XFEL Project Group, *The European X-Ray Free Electron Laser. Technical Design Report* (Hamburg: DESY, 2007), 23–40.
70. See S. Schreiber, *Status of the TTF FEL*, paper presented at the eleventh Workshop on RF Superconductivity, September, 8–12, 2003 at Travemünde/Lübeck, Germany.
71. See DESY, *A VUV Free Electron Laser at the TESLA Test Facility at DESY: Conceptual Design Report* (Hamburg: DESY, 1995).
72. See DESY, *A VUV Free Electron Laser*, 77–82.
73. J. Andruszkow et al., "First Observation of Self-Amplified Spontaneous Emission in a Free-Electron Laser at 109 nm Wavelength," *Physical Review Letters* 85, no. 18 (2000).
74. See Lohrmann and Söding, *Von schnellen Teilchen*, 250–251.
75. See, for example, TESLA Test Facility FEL Team, *SASE FEL at the TESLA Facility, Phase 2*, TESLA-FEL 2002-01 (Hamburg: DESY, 2002); Schneider, "FLASH."
76. On FLASH, see, for example, Heinze, Hallonsten, and Heinecke, "Turning the Ship"; Lohrmann and Söding, *Von schnellen Teilchen*, Chapter 13.8.
77. See TESLA Collaboration, *TESLA. The Superconducting Electron-Positron Linear Collider with an Integrated X-Ray Laser Laboratory: Technical Design Report* (Hamburg: TESLA Collaboration, 2001).
78. See H. Gutbrod et al., eds. *FAIR Baseline Technical Report: Executive Summary* (Darmstadt: Gesellschaft für Schwerionenphysik, 2006), 15–17.
79. See DESY, *Wissenschaftlicher Jahresbericht 2003* (Hamburg: DESY, 2004), 253.
80. See Wissenschaftsrat, *Stellungnahme zu neun Großgeräten*, 21.
81. See, for example, F. Neidhardt, "Institution, Organisation, Interaktion: Funktionsbedingungen des Wissenschaftsrats," *Leviathan* 40, no. 2 (2012); O. Bartz, *Der Wissenschaftsrat: Entwicklungslinien der Wissenschaftspolitik in der Bundesrepublik Deutschland 1957–2007* (Stuttgart: Franz Steiner Verlag, 2007).
82. These committees were *Großinvestitionen in der Grundlagenforschung* in 1975, also named *Maier-Leibnitz Committee* after its chairman Heinz Maier-Leibnitz, and *Großprojekte der Grundlagenforschung* in 1981, also named *Pinkau Committee* after its chairman Klaus Pinkau.
83. See, for example, Wissenschaftsrat, *Stellungnahme zu neun Großgeräten*, 89–91; H. Schunck, "Bundesweite Förderung der Physik in Deutschland," in *Physik im 21. Jahrhundert: Essays zum Stand der Physik*, ed. W. Martienssen and D. Röß (Berlin: Springer, 2011), 386.
84. Wissenschaftsrat, *Stellungnahme zu neun Großgeräten*.
85. See Wissenschaftsrat, *Stellungnahme zu neun Großgeräten*, 111–117.

86. TESLA Collaboration, *TESLA XFEL First Stage of the X-Ray Laser Laboratory: Technical Design Report Supplement* (Hamburg: DESY, 2002).
87. See Wissenschaftsrat, *Planungssicherheit für die Wissenschaft zur Entscheidung des Bundes über Großgeräte der Naturwissenschaftlichen Grundlagenforschung*, News Release, February 5, 2003; Gutbrod et al., *FAIR Baseline Technical Report*, 6.
88. Heinze, Hallonsten, and Heinecke, "Turning the Ship", 440.
89. See, for example, Meusel, E.-J. "Die Förderung der Großforschung durch das BMBF." In *Das Wissensministerium: Ein halbes Jahrhundert Forschungs- und Bildungspolitik in Deutschland*, eds. P. Weingart and N. Taubert (Weilerswist: Velbrück Wissenschaft, 2006), Section 1: Rechtsgrundladen; Stucke, A. "Staatliche Akteure der Wissenschaftspolitik." In *Handbuch Wissenschaftspolitik*, eds. S. Hornbostel, A. Knie and D. Simon (Wiesbaden: Verlag für Sozialwissenschaften, 2010).
90. See *Niederschrift über die Gemeinsame Konferenz der Kultus- und Finanzminister (-senatoren) der Länder der Bundesrepublik am 6. Februar 1958 in Bonn*, Akten des Sekretariats der Ständigen Konferenz der Kultusminister der Länder in der Bundesrepublik Deutschland, B 304/ 1731 2, Bundesarchiv; *Letter from Bundesminister für Atomenergie- und Wasserwirtschaft to Ministerpräsident des Landes Baden-Württemberg*, 15 March 1958, Akten des Sekretariats der Ständigen Konferenz der Kultusminister der Länder in der Bundesrepublik Deutschland, B 304/1731 2, Bundesarchiv.
91. See *Auszug aus der Niederschrift über die Sitzung Nr. 32 des Hochschulausschusses der Kultusministerkonferenz am 3./4. Mai 1957 in Bonn*, Akten des Sekretariats der Ständigen Konferenz der Kultusminister der Länder in der Bundesrepublik Deutschland, B 304/ 1731 2, Bundesarchiv.
92. *Rahmenvereinbarung zwischen Bund und Ländern über die gemeinsame Förderung der Forschung nach Artikel 91b GG*, November 28, 1975.
93. See, for example, Wissenschaftsrat, *Systemevaluation der HGF: Stellungnahme des Wissenschaftsrates zur Hermann von Helmholtz-Gemeinschaft Deutscher Forschungszentren* (Berlin: Wissenschaftsrat, 2001), 12; D. Hoffmann and H. Trischler, "Die Helmholtz-Gemeinschaft in Historischer Perspektive," in *20 Jahre Helmholtz-Gemeinschaft*, eds. J. Mlynek and A. Bittner (Bonn: Helmholtz Gemeinschaft, 2015), 15; M. Kölbel, "Das Bundesministerium für Bildung und Forschung (BMBF) als Wissenschaftspolitischer Akteur," in *Handbuch Wissenschaftspolitik*, eds. S. Hornbostel, A. Knie and D. Simon (Wiesbaden: Verlag für Sozialwissenschaften, 2010), 534.

94. As of 2019, the Helmholtz Association includes eighteen research centers: Alfred-Wegener-Institut Helmholtz-Zentrum für Polar- und Meeresforschung, Deutsches Elektronen-Synchrotron (DESY), Deutsches Zentrum für Luft- und Raumfahrt (DLR), Deutsches Krebsforschungszentrum, Deutsches Zentrum für Neurodegenerative Erkrankungen (DZNE), Forschungszentrum Jülich, GEOMAR Helmholtz-Zentrum für Ozeanforschung Kiel, GSI Helmholtzzentrum für Schwerionenforschung, Helmholtz-Zentrum Berlin für Materialien und Energie, Helmholtz-Zentrum Dresden-Rossendorf, Helmholtz-Zentrum für Infektionsforschung, Helmholtz-Zentrum für Umweltforschung – UFZ, Helmholtz-Zentrum Geesthacht Zentrum für Material- und Küstenforschung, Helmholtz Zentrum München – Deutsches Forschungszentrum für Gesundheit und Umwelt, Helmholtz-Zentrum Potsdam – Deutsches GeoForschungsZentrum (GFZ), Karlsruher Institut für Technologie, Max-Delbrück-Centrum für Molekulare Medizin in der Helmholtz-Gemeinschaft, Max-Planck-Institut für Plasmaphysik, CISPA – Helmholtz-Zentrum für Informationssicherheit.
95. See, for example, Hoffmann and Trischler, "Die Helmholtz-Gemeinschaft"; S. Helling-Moegen, *Forschen nach Programm. Die programmorientierte Förderung in der Helmholtz-Gemeinschaft: Anatomie einer Reform* (Marburg: Tectum, 2009); T. Heinze and N. Arnold, "Governanceregimes im Wandel," *KZfSS Kölner Zeitschrift für Soziologie und Sozialpsychologie* 60, no. 4 (2008).
96. See Heinze, Hallonsten, and Heinecke, "Turning the Ship", 431.
97. See Satzung des Vereins "Hermann von Helmholtz-Gemeinschaft Deutscher Forschungszentren e.V.", §8–10.
98. See, for example, Heinze and Arnold, "Governanceregimes im Wandel", 702; Hallonsten and Heinze, "Institutional Persistence", 456.
99. Heinze, Hallonsten, and Heinecke, "Turning the Ship", 431.
100. Correspondence with Hermann Schunck (June 15, 2017). Translated from German.
101. See Heinze, Hallonsten, and Heinecke, "Turning the Ship", 439.
102. See Heinze, Hallonsten, and Heinecke, "Turning the Ship", 439.
103. See *Niederschrift über die 98. Sitzung des Verwaltungsrates am 12./13. Juni 2003 in Zeuthen*, DESY Archive.
104. See DESY, *Worldwide Consensus of Particle Physicists: Superconductivity for Future Particle Accelerator Project ILC*, News Release (August 20, 2004).
105. See Interview with Karl Witte (September 22, 2015, Halstenbek); *Bericht des Direktoriums zur 102. Sitzung des DESY Verwaltungsrates am 2./3. Dezember 2004 bei DESY in Hamburg*, DESY Archive.

106. See Schleswig-Holsteinischer Landtag, *Gesetzentwurf der Landesregierung zu dem Abkommen zwischen der Bundesrepublik Deutschland, der Freien und Hansestadt Hamburg und dem Land Schleswig-Holstein zum Bau und Betrieb der Europäischen Freie-Elektronen-Röntgenlaseranlage (XFEL) und dem Staatsvertrag zwischen der Freien und Hansestadt Hamburg und dem Land Schleswig-Holstein über den Bau und Betrieb der Europäischen Freie-Elektronen-Röntgenlaseranlage (XFEL)*, January 15, 2010.
107. See *Staatsvertrag zwischen der Freien und Hansestadt Hamburg und dem Land Schleswig-Holstein über die Schaffung der planerischen Voraussetzungen für die Errichtung und den Betrieb eines Freie Elektronen Lasers im Röntgenlaserbereich*, September 28, 2004.
108. See DESY, *Wissenschaftlicher Jahresbericht 2003*, 127.
109. See *Final Act of the Conference of Plenipotentiaries for the Establishment of a European X-Ray Free-Electron Laser Facility*, 30 November 2009, §1.
110. See DESY, *Wissenschaftlicher Jahresbericht 2004* (Hamburg: DESY, 2005), 200.
111. See *Final Report Summary – PRE-XFEL* (Preparatory Activities for the Implementation of the European X-Ray Free-Electron Laser Facility), Grant agreement ID: 211604, March 1, 2013.
112. See *Final Act of the Conference of Plenipotentiaries for the Establishment of a European X-Ray Free-Electron Laser Facility*, November 30, 2009.
113. See Art. 1 and 5 of the *European XFEL Convention*.
114. Correspondence with Hermann Schunck (September 30, 2017). Translated from German.
115. Correspondence with Hermann Schunck (September 30, 2017 and March 9, 2020).
116. H. Neal, T. Smith, and J. McCormick, *Beyond Sputnik: U.S. Science Policy in the Twenty-First Century* (Ann Arbor: University of Michigan Press, 2008), 206.
117. Correspondence with Herrmann Schunck (September 30, 2017).
118. See Heinze, Hallonsten, and Heinecke, "Turning the Ship", 433.
119. Correspondence with Hermann Schunck (September 30, 2017).
120. "Memorandum Submitted by the Head of School of Optometry (FC 22)." In *The Impact of Spending Cuts on Science and Scientific Research*, House of Commons/ Science and Technology Committee, Sixth Report of Session 2009–10 (London: House of Commons, 2010), ev125–ev126.
121. See House of Commons/ Science and Technology Committee, *The Impact of Spending*, ev61; House of Commons/ Science and Technology Committee, *Astronomy and Particle Physics*, Fourth Report of Session 2010–12 (London: House of Commons, 2011), 5–6.

122. See House of Commons/ Science and Technology Committee, *Astronomy and Particle Physics*, 5.
123. House of Commons Science and Technology, *The Impact of Spending*, 20.
124. See, for example, J. Womersley, *The STFC Science Programme 2010–15*, presentation given at the STFC Town Meeting 2009; House of Commons Science and Technology, *The Impact of Spending*, ev12, ev126; European XFEL, *United Kingdom Withdraws Participation in European XFEL*, News Release (December 18, 2009).
125. See Science Board, *Minutes of the Science Board's Meeting* (December 7–8, 2009).
126. See, for instance, European XFEL, *UK Contributes a High-Energy Optical Laser to HED Instrument*, News Release (July 21, 2015); STFC, *UK Joins World Leading X-Ray Laser Facility*, News Release (March 19, 2018).
127. See Medical Research Council, *UK invests in World's Most Advanced Crystallography Facility*, News Release (June 3, 2014).
128. See M. Banks, "Funding Shortfall Hits European X-ray Laser Facility," *Physics World* (July 29, 2011).
129. See European XFEL, *The United Kingdom Officially Joins European XFEL*, News Release (March 19, 2018).
130. Interview with Hermann Schunck (June 15, 2017).
131. See H. Schunck, *Meine Jahre mit DESY*.
132. See *Memorandum of Understanding Between the Deutsches Elektronen-Synchrotron (DESY) and the Stanford Linear Accelerator Center (SLAC) Establishing a Collaborative Research Effort to Enable the Exploitation and Expansion of the Scientific Capabilities of The Linac Coherent Light Source and The TESLA X-Ray Free-Electron Laser.*
133. Interview with Hermann Schunck (June 15, 2017).
134. See A. Bonucci, *In-Kind Contributions to the European XFEL Facility*, presentation given at the second In-Kind Contributions Workshop (IKCW) at FAIR, 2015.
135. See "in-kind contributions" at European XFEL's website: https://www.xfel.eu/facility/construction/in_kind_contributions/index_eng.html
136. See, for example, Bonucci, *In-Kind Contributions;* K. Witte, *Comparison of Legal Difficulties/Advantages of ESRF and European XFEL*, presentation given at the second meeting of the CTA Legal Working Group, Paris, 2012; S. Prat, *Management and Control of In-Kind Contributions. Case Study: European XFEL Facility*, presentation given at RAMIRI 2012 Learning Programme, Trieste, 2012.
137. See European XFEL GmbH, *Annual Report 2016* (Schenefeld: European XFEL GmbH, 2017), 71.
138. See Art. 5.4 and 6.3 of the *European XFEL Convention.*

139. See European Commission, *Review of the S&T Cooperation Agreement between the European Union and Russia* (Brussels: European Commission, 2013).
140. See OECD data on gross domestic spending on R&D. Online available: https://data.oecd.org, last accessed September 5, 2017.
141. See, for example, L. Graham and I. Dezhina, *Science in the New Russia: Crisis, Aid, Reform* (Bloomington: Indiana University Press, 2008), 18–19; S. Radosevic, "Patterns of Preservation, Restructuring and Survival: Science and Technology Policy in Russia in Post-Soviet Era." *Research Policy*, no. 3 (2003).
142. See I. Yegorov, "Post-Soviet Science: Difficulties in the Transformation of the R&D Systems in Russia and Ukraine," *Research Policy* 38, no. 4 (2009).
143. See, for example, V. Kessler, *Die sichtbare Hand des Staates: Eine rechtsökonomische Analyse russischer Staatskorporationen* (Berlin: BWV Berliner Wissenschafts-Verlag, 2014), 231–235, 260–266.
144. See, for example, R. Götz, "Deutschland und Russland – 'strategische Partner'?" *APuZ (Aus Politik und Zeitgeschichte)* (March 2006); S. Heinecke, *Die deutsche Russlandpolitik 1991–2005: Entwicklungen und Gesellschaftliche Einflüsse in außenpolitischen Entscheidungsprozessen* (Doctoral Thesis, Cologne, 2011).
145. R. Sakwa, "Vladimir Putin and Russian Foreign Policy Towards the West: Towards a New Realism." In *Russia and Europe in the Twenty-First Century: An Uneasy Partnership*, eds. J. Gower and G. Timmins (London: Anthem Press, 2009), 3–4.
146. See, for example, T. Forsberg, "From Ostpolitik to Frostpolitik? Merkel, Putin and German Foreign Policy Towards Russia," *International Affairs* 92, no. 1 (2016); G. Timmins and J. Gower, "Introduction: Russia and Europe: What Kind of Partnership?" In *Russia and Europe in the Twenty-First Century: An Uneasy Partnership, eds.* J. Gower and G. Timmins (London: Anthem Press, 2009).
147. See R. Götz, "Deutschland und Russland – 'strategische Partner'?" *APuZ (Aus Politik und Zeitgeschichte)* (March 2006), 15; S. Heinecke, *Die deutsche Russlandpolitik 1991–2005: Entwicklungen und Gesellschaftliche Einflüsse in außenpolitischen Entscheidungsprozessen* (Doctoral Thesis, Cologne, 2011), 114.
148. See G. Gorzka, *Knowledge Transfer between Germany and Russia: Drivers, Instruments and Impact* (Kassel: Kassel University Press, 2016), 14–15.
149. See Gorzka, *Knowledge Transfer*, 25–29.
150. See European Commission, *Roadmap for EU-Russia S&T Cooperation* (Brussels: European Commission, 2017).
151. See European Commission, *Roadmap EU-Russia*, 20–40.

152. *Gemeinsame Erklärung über die strategische Partnerschaft auf dem Gebiet der Bildung, Forschung und Innovation zwischen der Bundesrepublik Deutschland und der Russischen Föderation* (April 11, 2005).
153. Gorzka, *Knowledge Transfer*, 14–15.
154. Correspondence with Rainer Koepke (October 3, 2017).
155. Correspondence with Rainer Koepke (October 3, 2017).
156. Correspondence with Rainer Koepke (October 3, 2017).
157. See, for example, H. Adomeit, "German-Russian Relations: Balance Sheet since 2000 and Perspectives until 2025", *ifri. Comité d'Études des Relations Franco-Allemandes*, 2012, 6; J. Yoder, "From Amity to Enmity: German-Russian Relations in the Post Cold War Period." *German Politics and Society* 33, no. 3 (2015), 56.
158. Quoted in O. Hallonsten, "The Politics of European Collaboration in Big Science," in *The Global Politics of Science and Technology – Vol. 2*, eds. M. Mayer, M. Carpes, and R. Knoblich (Berlin, Heidelberg: Springer, 2014), 40.
159. DESY/ IRI Hamburg Office, *Ioffe-Röntgen-Institut (IRI)*, (without date, unpublished).
160. See M. Karaulova et al., "Science System Path-Dependencies and their Influences: Nanotechnology Research in Russia," *Scientometrics* 107 (2016).
161. See, for example, C. Kehrt, *Mit Molekülen spielen: Wissenschaftskulturen der Nanotechnologie zwischen Politik und Medien* (Bielefeld: transcript Verlag, 2016); J. Schummer, *Nanotechnologie – Spiele mit Grenzen* (Frankfurt a.M.: Suhrkamp, 2009).
162. Quoted in F. Westerlund, *Russian Nanotechnology R&D. Thinking Big About Small Scale Science* (Stockholm: Swedish Defence Research Agency, 2011), 35.
163. See, for example, Q. Schiermeier, "High Hopes for Russia's Nanotech Firms," *Nature* 461, no. 7267 (2009); Q. Schiermeier, "Russia Pins its Hopes on 'Nano'," *Nature* 448, no. 7151 (2007).
164. See, for example, A. Zubkovskaya and S. Michailova, "The Development of Russian Multinational Enterprises from the 1990s to the Present," *Organizations and Markets in Emerging Economies* 5, no. 2 (2014), 64; Kessler, *Die sichtbare Hand des Staates,* 135–136; F. Westerlund, *Russian Nanotechnology*, 47–52.
165. See M. Karaulova et al., "Science System" (2016); Q. Schiermeier, "High Hopes" (2009).
166. See F. Westerlund, *Russian Nanotechnology R&D,* 39.
167. See W. McCray, "'Globalization with Hardware': ITER's Fusion of Technology, Policy, and Politics," *History and Technology* 26, no. 4 (2010), 295.

168. See S. Smirnov, "Russia–CERN Cooperation: Current Status and Perspectives," *Journal of Physics: Conference Series*, 1406, no. 1 (2019).
169. See *Prime Minister Vladimir Putin holds a session of the Government Commission on High Technology and Innovation in Dubna* (July 5, 2011), Archive of the Official Site of the 2008–2012 Prime Minister of the Russian Federation Vladimir Putin.
170. CREMLIN, First CREMLIN Recommendations for the European-Russian Megascience Collaboration (June 5, 2018, Hamburg: DESY).
171. *Final Report Summary: PRE-XFEL.*
172. For example, EIROforum (European Intergovernmental Research Organisations), EUCALL (European Cluster of Advanced Laser Light Sources) FELs of Europe or LEAPS (League of European Accelerator-based Photon Sources).
173. Interview with Karl Witte (September 22, 2015, Halstenbek).
174. See Art 12.3 of the *Articles of Association of the European XFEL.*
175. See "Declaration by the Government of the Russian Federation regarding the participation of the Russian Federation in the project to build and operate a European X-ray Free-Electron Laser Facility." In *Final Act of the Conference of Plenipotentiaries of the European XFEL.*
176. See "Declaration by the Government of the Russian Federation regarding the participation of the Russian Federation in the project to build and operate a European X-ray Free-Electron Laser Facility." In *Final Act of the Conference of Plenipotentiaries of the European XFEL.*
177. See, for example, Hallonsten, *Big Science Transformed*, 116ff; Hallonsten, "The Politics of European Collaboration", 44.
178. See, for example, Hallonsten, *Big Science Transformed*, 116ff; Hallonsten, "The Politics of European Collaboration", 44.
179. See Art. 6 of the *European XFEL Convention.*
180. See "Declaration by the Government of the Russian Federation regarding the participation of the Russian Federation in the project to build and operate a European X-ray Free-Electron Laser Facility." In *Final Act of the Conference of Plenipotentiaries of the European XFEL.*
181. See RUSNANO, *RUSNANO Takes Part in Construction and Operation of the Most Powerful in the World X-Ray Free Electron Laser*, News Release (April 16, 2009).
182. Differing view: C. Harringa, "Zwischen Völkerrecht und Frascati."
183. See, for example, DESY, *Memorandum on the Realisation of the First Stage of the European TESLA XFEL LABORATORY, 24 October 2002*, DESY Archive.
184. See, for example, Art 1.3 of the *European XFEL Convention.*

Bibliography

Bundesarchiv, Koblenz, Germany

B 304/1731 2: Akten des Sekretariats der Ständigen Konferenz der Kultusminister der Länder in der Bundesrepublik Deutschland.

DESY Archive, Hamburg, Germany (Chronological Order)

DESY. *Planungsschritte Linearcollider.*

TESLA Collaboration. *TTF Collaboration Meeting – Final Report*, June 1–2 1993, Hamburg.

Niederschrift über die 78. Sitzung des Verwaltungsrates am 13. Juli 1993 in Zeuthen.

Niederschrift über die 82. Sitzung des Verwaltungsrates am 28. Juni 1995 in Zeuthen.

Niederschrift über die 87. Sitzung des Verwaltungsrates am 12. November 1997 in Zeuthen.

Niederschrift über die 95. Sitzung des Verwaltungsrates am 5. Dezember 2001 in Hamburg.

DESY. *Memorandum on the Realisation of the First Stage of the European TESLA XFEL LABORATORY*, 24 October 2002.

Niederschrift über die 98. Sitzung des Verwaltungsrates am 12./13. Juni 2003 in Zeuthen.

100. Sitzung des DESY Verwaltungsrates am 24./25. Juni 2004 bei DESY in Zeuthen.

Bericht des Direktoriums zur 102. Sitzung des DESY Verwaltungsrates am 2./3. Dezember 2004 bei DESY in Hamburg.

Legal/Governmental Documents (Chronological Order)

Bürgerschaft der Freien und Hansestadt Hamburg, *Mitteilung des Senats an die Bürgerschaft. Gesetz zum Staatsvertrag zwischen der Freien und Hansestadt Hamburg und dem Land Schleswig-Holstein über die Schaffung der planerischen Voraussetzungen für die Errichtung und den Betrieb eines Linearbeschleunigers*, Drucksache 16/436, 4 November 1998.

Bundestag, *Antrag der Abgeordneten Ulrike Flach et al. und der Fraktion der F.D.P.: Forschungsstandort Deutschland stärken – Zukunftsprojekt „TESLA" nicht gefährden*, Drucksache 14/4646, 15 November 2000.

Memorandum of Understanding Between the Deutsches Elektronen-Synchrotron (DESY) and the Stanford Linear Accelerator Center (SLAC) Establishing a

Collaborative Research Effort to Enable the Exploitation and Expansion of the Scientific Capabilities of The Linac Coherent Light Source and The TESLA X-Ray Free-Electron Laser, 1 November 2002.

Staatsvertrag zwischen der Freien und Hansestadt Hamburg und dem Land Schleswig-Holstein über die Schaffung der planerischen Voraussetzungen für die Errichtung und den Betrieb eines Freie Elektronen Lasers im Röntgenlaserbereich, 28 September 2004.

Gemeinsame Erklärung über die strategische Partnerschaft auf dem Gebiet der Bildung, Forschung und Innovation zwischen der Bundesrepublik Deutschland und der Russischen Föderation, 11 April 2005.

Agreement on the Establishment of the ITER International Fusion Energy Organization for the Joint Implementation of the ITER Project, 21 November 2006.

Articles of Association of the "European X-Ray Free-Electron Laser Facility GmbH" (European XFEL GmbH), 30 November 2009.

Convention Concerning the Construction and Operation of a European X-Ray Free-Electron Laser Facility, 30 November 2009.

Final Act of the Conference of Plenipotentiaries for the Establishment of a European X-Ray Free-Electron Laser Facility, 30 November 2009.

Convention concerning the Construction and Operation of a Facility for Antiproton and Ion Research in Europe (FAIR), 12 October 2010.

Schleswig-Holsteinischer Landtag, *Gesetzentwurf der Landesregierung zu dem Abkommen zwischen der Bundesrepublik Deutschland, der Freien und Hansestadt Hamburg und dem Land Schleswig-Holstein zum Bau und Betrieb der Europäischen Freie-Elektronen-Röntgenlaseranlage (XFEL) und dem Staatsvertrag zwischen der Freien und Hansestadt Hamburg und dem Land Schleswig-Holstein über den Bau und Betrieb der Europäischen Freie-Elektronen-Röntgenlaseranlage (XFEL)*, 15 January 2010.

Letter of Intent of the Deutsches Elektronen-Synchrotron DESY (Hamburg, Germany) and the National Research Centre "Kurchatov Institute" (Moscow, Russian Federation) Concerning the German-Russian Cooperation in Strategic Research Fields: The Ioffe-Röntgen Institute, 23 May 2011.

Protocole d'Adhésion du Gouvernment de la Féderation de Russie à la Convention du 16 Décembre 1988 Relative à la Construction et à l'Exploitation d'une Installation Européenee de Rayonnement de Synchrotron, 23 June 2014.

Satzung des Vereins "Hermann von Helmholtz-Gemeinschaft Deutscher Forschungszentren e.V." (Von der Gründungsversammlung am 12. September 2001 festgestellte Fassung mit den von der Mitgliederversammlung auf ihrer 79. Sitzung am 15. Oktober 2003, auf ihrer 87. Sitzung am 13. September 2006, auf ihrer 99. Sitzung am 19. September 2012 und auf ihrer 111. Sitzung am 11. September 2018 beschlossenen Änderungen, eingetragen in das Vereinsregister unter VR 7942 beim Amtsgericht Bonn am 17. Oktober 2018), 11 September 2018.

References

Adomeit, H. "German-Russian Relations: Balance Sheet since 2000 and Perspectives until 2025." *ifri. Comité d'Études des Relations Franco-Allemandes*, 2012.

Andruszkow, J., Aune, B., Ayvazyan, V., Baboi, N., Bakker, R., Balakin, V., Barni, D., et al. "First Observation of Self-Amplified Spontaneous Emission in a Free-Electron Laser at 109 nm Wavelength." *Physical Review Letters* 85, no. 18 (2000): 3825–3829.

Banks, M. "Funding Shortfall Hits European X-ray Laser Facility." *Physics World* (29 July 2011). Online available: http://physicsworld.com/cws/article/news/2011/jul/29/funding-shortfall-hits-european-x-ray-laser-facility, last accessed 20 May 2019.

Bartz, O. *Der Wissenschaftsrat: Entwicklungslinien der Wissenschaftspolitik in der Bundesrepublik Deutschland 1957–2007* (Stuttgart: Franz Steiner Verlag, 2007).

Behnke, T., Brau, J., Foster, B., Fuster, B., Harrison, M., McEwan Paterson, J., Peskin, M., Stanitzki, M., Walker, N., and Yamamoto, H. (eds.) *The International Linear Collider: Technical Design Report.* 2013.

Bonucci, A. *In-Kind Contributions to the European XFEL Facility*, presentation given at the 2nd In-Kind Contributions Workshop (IKCW 2015) at FAIR. Online available: https://indico.gsi.de/event/3426/, last accessed 1 May 2019.

Brinkmann, R., Materlik, G., Rossbach, J., and Wagner, A. (eds.) *Conceptual Design Report of a 500 GeV e^+e^- Linear Collider with Integrated X-Ray Laser Facility.* Hamburg: DESY, 1997.

Brinkmann, R., Schmüser, P., and Wiik, B. "TESLA." *Physik in unserer Zeit* 30, no. 2 (1999): 62–68.

Castells, M. and Cardoso, G. (eds.) *The Network Society: From Knowledge to Policy.* Washington DC: Johns Hopkins Center for Transatlantic Relations, 2005.

CREMLIN. *First CREMLIN Recommendations for the European-Russian Megascience Collaboration.* Hamburg: DESY, 5 June 2018.

DESY XFEL Project Group. *The European X-Ray Free Electron Laser. Technical Design Report.* Hamburg: DESY XFEL Project Group/ European XFEL Project Team, 2007.

DESY. *Particle Physics 2016: Jahresbericht des Bereichs Teilchenphysik bei DESY.* Online available: http://www.desy.de/sites2009/site_www-desy/content/e410/e84441/e234837/ParticlePhysics2016_ger.pdf, last accessed 10 November 2017.

DESY. *Wissenschaftlicher Jahresbericht 2004.* Hamburg: DESY, 2005.

DESY. *Worldwide Consensus of Particle Physicists: Superconductivity for Future Particle Accelerator Project ILC*, News Release, 20 August 2004. Online available: https://www.desy.de/information__services/press/press_releases/2004/pr_200804/index_eng.html, last accessed 20 March 2020.

DESY. *Wissenschaftlicher Jahresbericht 2003*. Hamburg: DESY, 2004.
DESY. *Wissenschaftlicher Jahresbericht 1998*. Hamburg: DESY, 1999.
DESY. *Wissenschaftlicher Jahresbericht 1995*. Hamburg: DESY, 1996.
DESY. *A VUV Free Electron Laser at the TESLA Test Facility at DESY: Conceptual Design Report*. Hamburg: DESY, 1995.
Elzinga, A. "Features of the Current Science Policy Regime: Viewed in Historical Perspective." *Science and Public Policy* 39, no. 4 (2012): 416–428.
European Commission. *Roadmap for EU-Russia S&T Cooperation*. Brussels: European Commission, 2017.
European Commission. *Review of the S&T Cooperation Agreement between the European Union and Russia*. Brussels: European Commission, 2013.
European Committee for Future Accelerators. *Report 1967*. Geneva: CERN, 1967.
European XFEL. *Annual Report 2016*. Schenefeld: European XFEL GmbH, 2017.
European XFEL. *The United Kingdom Officially Joins European XFEL*, News Release, 19 March 2018. Online available: https://www.xfel.eu/news_and_events/news/index_eng.html?openDirectAnchor=1435&two_columns=0, last accessed 25 March 2018.
European XFEL. *UK Contributes a High-Energy Optical Laser to HED Instrument*, News Release 21 July 2015. Online available: https://www.xfel.eu/news_and_events/news/index_eng.html?openDirectAnchor=1182&two_columns=0, last accessed 10 March 2018.
European XFEL. *United Kingdom Withdraws Participation in European XFEL*, News Release 18 December 2009. Online available: https://www.xfel.eu/news_and_events/archived_news/2009/20091218/index_eng.html, last accessed 25 April 2017.
Forsberg, T. "From Ostpolitik to Frostpolitik? Merkel, Putin and German Foreign Policy towards Russia." *International Affairs* 92, no. 1 (2016): 21–42.
Galison, P. and Hevly B. (eds) *Big Science: The Growth of Large-Scale Research* (Stanford: Stanford University Press, 1992).
Gorzka, G. *Knowledge Transfer between Germany and Russia: Drivers, Instruments and Impact*. Kassel: Kassel University Press, 2016.
Götz, R. "Deutschland und Russland – ‚strategische Partner'?" *APuZ (Aus Politik und Zeitgeschichte)* (March 2006): 14–23.
Graham, L. and Dezhina, I. *Science in the New Russia: Crisis, Aid, Reform*. Bloomington: Indiana University Press, 2008.
Greenberg, D. *The Politics of Pure Science*, new edition. Chicago: University of Chicago Press, 1999.
Gutbrod, H., Augustin, I., Eickhoff, H., Groß, K.-D., Henning, W., Krämer, D., and Walter G. (eds.) *FAIR Baseline Technical Report: Executive Summary*. Darmstadt: Gesellschaft für Schwerionenphysik, 2006.
Habfast, C. *Grossforschung mit kleinen Teilchen: Das Deutsche Elektronen-Synchrotron, DESY, 1956–1970*. Berlin, New York: Springer, 1989.

Hallonsten, O. *Big Science Transformed: Science, Politics and Organization in Europe and the United States.* Cham: Palgrave Macmillan, 2016.

Hallonsten, O. "The Politics of European Collaboration in Big Science." In *The Global Politics of Science and Technology – Vol. 2.* Edited by M. Mayer, M. Carpes and R. Knoblich, 31–46. Berlin, Heidelberg: Springer, 2014.

Hallonsten, O. "Continuity and Change in the Politics of European Scientific Collaboration." *Journal of Contemporary European Research* 8, no. 3 (2012): 300–319.

Hallonsten, O. and Heinze, T. "Institutional Persistence Through Gradual Organizational Adaptation: Analysis of National Laboratories in the USA and Germany." *Science and Public Policy* 39, no. 4 (2012): 450–463.

Harringa, C. "Zwischen Völkerrecht und Frascati: Praktische Aspekte der Ausgestaltung internationaler Kooperationen am Deutschen Elektronen Synchrotron (DESY)." *Ordnung der Wissenschaft* 2 (2018): 69–74.

Heinecke, S. *Die deutsche Russlandpolitik 1991–2005: Entwicklungen und Gesellschaftliche Einflüsse in außenpolitischen Entscheidungsprozessen.* Doctoral Thesis, University of Cologne, 2011.

Heinze, T., and Arnold, N. "Governanceregimes im Wandel." *KZfSS Kölner Zeitschrift für Soziologie und Sozialpsychologie* 60, no. 4 (2008): 686–722.

Heinze, T., Hallonsten, O., and Heinecke, S. "Turning the Ship: The Transformation of DESY, 1993–2009." *Physics in Perspective* 19, no. 4 (2017): 424–451.

Heinze, T., Hallonsten, O., and Heinecke, S. "From Periphery to Center: Synchrotron Radiation at DESY, Part II: 1977–1993." *Historical Studies in the Natural Sciences* 45, no. 4 (2015a): 513–548.

Heinze, T., Hallonsten, O., and Heinecke, S. "From Periphery to Center: Synchrotron Radiation at DESY, Part I: 1962–1977." *Historical Studies in the Natural Sciences* 45, no. 3 (2015b): 447–492.

Helling-Moegen, S. *Forschen nach Programm: Die programmorientierte Förderung in der Helmholtz-Gemeinschaft: Anatomie einer Reform.* Marburg: Tectum, 2009.

Hoddeson, L. "The First Large-Scale Application of Superconductivity: The Fermilab Energy Doubler, 1972–1983." Historical Studies in the Physical and Biological Sciences 18, no. 1 (1987): 25–54.

Hoddeson, L., Brown, L. M., Dresden, M., and Riordan, M. (eds.) *The Rise of the Standard Model: A History of Particle Physics from 1964 to 1979.* Cambridge: Cambridge University Press, 1997;

Hoffmann, D. and Trischler, H. "Die Helmholtz-Gemeinschaft in Historischer Perspektive." In *20 Jahre Helmholtz-Gemeinschaft.* Edited by J. Mlynek and A. Bittner, 9–47. Bonn: Helmholtz Gemeinschaft, 2015.

House of Commons/ Science and Technology Committee. *The Impact of Spending Cuts on Science and Scientific Research* (Sixth Report of Session 2009–10). London: House of Commons, 2010.

Kallerud, E., Klitkou, A., Sutherland Olsen, D., Scordato, L., Amanatidou, E., Upham, P., Nieminen, M., Lima-Toivanen, M. and Oksanen, J. *Dimensions of Research and Innovation Policies to Address Grand and Global Challenges.* NIFU Working Paper 13, 2013.

Karaulova, M., Gök, A., Shackleton, O., and Shapira, P. "Science System Path-Dependencies and their Influences: Nanotechnology Research in Russia." *Scientometrics* 107 (2016): 645–670.

Kehrt, C. *Mit Molekülen spielen: Wissenschaftskulturen der Nanotechnologie zwischen Politik und Medien* (Bielefeld: transcript Verlag, 2016).

Kessler, V. *Die sichtbare Hand des Staates: Eine rechtsökonomische Analyse russischer Staatskorporationen.* Berlin: BWV Berliner Wissenschafts-Verlag, 2014.

Kevles, D. "Good-Bye to the SSC: On the Life and Death of the Superconducting Super Collider." *Engineering & Science* 58, no. 2 (1995): 16–25.

Kölbel, M. "Das Bundesministerium für Bildung und Forschung (BMBF) als Wissenschaftspolitischer Akteur." In *Handbuch Wissenschaftspolitik.* Edited by S. Hornbostel, A. Knie and D. Simon, 533–548. Wiesbaden: Verlag für Sozialwissenschaften, 2010.

Krige, J. *American Hegemony and the Postwar Reconstruction of Science in Europe.* Cambridge, MA: MIT Press, 2006.

Kunz, C. *Synchrotronstrahlung bei DESY: Anfänge.* Private print, 2012.

Lengwiler, M. "Kontinuitäten und Umbrüche in der Deutschen Wissenschaftspolitik des 20. Jahrhunderts." In *Handbuch Wissenschaftspolitik.* Edited by S. Hornbostel, A. Knie and D. Simon, 6–18. Wiesbaden: Verlag für Sozialwissenschaften, 2010.

Lohrmann, E. and Söding, P. *Von schnellen Teilchen und hellem Licht: 50 Jahre Deutsches Elektronen-Synchrotron DESY.* Weinheim: Wiley, 2009.

Lohrmann, E. and Söding, P. "DESY Marks 50 Years of Accelerator Research." *CERN Courier*, 7 December 2009. Online available: https://cerncourier.com/a/desy-marks-50-years-of-accelerator-research/, last accessed 10 March 2020.

Maiani, L. and Bonolis, L. "The LHC Timeline: A Personal Recollection (1980–2012)." *The European Physical Journal H* 42, 4–5 (2017): 475–505.

Materlik, G. and Tschentscher, T. (eds.) *TESLA Technical Design Report: Part V: The XFEL.* Hamburg: DESY, 2001.

McCray, W. "'Globalization with Hardware': ITER's Fusion of Technology, Policy, and Politics." *History and Technology* 26, no. 4 (2010): 283–312.

Medical Research Council. *UK invests in World's Most Advanced Crystallography Facility*, News Release, 3 June 2014. Online available: https://mrc.ukri.org/news/browse/uk-invests-in-worlds-most-advanced-crystallography-facility, last accessed 10 April 2019.

Meusel, E.-J. "Die Förderung der Großforschung durch das BMBF." In *Das Wissensministerium: Ein halbes Jahrhundert Forschungs- und Bildungspolitik in Deutschland*. Edited by P. Weingart and N. Taubert, 144–157. Weilerswist: Velbrück Wissenschaft, 2006.

Neal, H., Smith, T., and McCormick, J. *Beyond Sputnik: U.S. Science Policy in the Twenty-First Century*. Ann Arbor: University of Michigan Press, 2008.

Neidhardt, F. "Institution, Organisation, Interaktion: Funktionsbedingungen des Wissenschaftsrats." *Leviathan* 40, no. 2 (2012): 271–296.

Nuhn, H.-D. "From Storage Rings to Free Electron Lasers for Hard X-Rays." *Journal of Physics: Condensed Matter* 16, no. 33 (2004): 13–21.

Oreskes, N. and Krige, J. (eds.) *Science and Technology in the Global Cold War*. Cambridge, MA: MIT Press, 2014.

Pellegrini, C. "The History of X-Ray Free-Electron Lasers." *The European Physical Journal H* 37, no. 5 (2012): 659–708.

Prat, S. *Management and Control of In-Kind Contributions. Case Study: European XFEL Facility*, presentation given at RAMIRI 2012 Learning Programme, Trieste, 2012.

Prime Minister of the Russian Federation Vladimir Putin, *Prime Minister Vladimir Putin holds a session of the Government Commission on High Technology and Innovation in Dubna*, 5 July 2011, Archive of the Official Site of the 2008–2012 Prime Minister of the Russian Federation Vladimir Putin. Online available: http://archive.premier.gov.ru/eng/events/news/15785/video.html, last accessed 10 March 2020.

Radosevic, S. "Patterns of Preservation, Restructuring and Survival: Science and Technology Policy in Russia in Post-Soviet Era." *Research Policy*, no. 3 (2003): 1105–1124.

Riordan, M. "A Bridge Too Far: The Demise of the Superconducting Super Collider." *Physics Today* 69, no. 10 (2016): 48–54.

Riordan, M., Hoddeson, L. and Kolb, A. *Tunnel Visions: The Rise and Fall of the Superconducting Super Collider*. Chicago: University of Chicago Press, 2015.

Ritter, G. *Großforschung und Staat in Deutschland: Ein historischer Überblick*. München: Beck, 1992.

RUSNANO. *RUSNANO Takes Part in Construction and Operation of the Most Powerful in the World X-Ray Free Electron Laser*, News Release, 16 April 2009. Online available: http://en.rusnano.com/press-centre/news/88397, last accessed 12 January 2020.

Sakwa, R. "Vladimir Putin and Russian Foreign Policy Towards the West: Towards a New Realism." In *Russia and Europe in the Twenty-First Century: An Uneasy Partnership*. Edited by J. Gower and G. Timmins, 1–22. London: Anthem Press, 2009.

Schiermeier, Q. "High Hopes for Russia's Nanotech Firms." *Nature* 461, no. 7267 (2009): 1036–1037.

Schiermeier, Q. "Russia Pins its Hopes on 'Nano'." *Nature* 448, no. 7151 (2007): 233.

Schneider, J. "FLASH – From Accelerator Test Facility to the First Single-Pass Soft X-Ray Free-Electron Laser." *Journal of Physics B* 43, no. 19 (2010).

Schreiber, S. *Status of the TTF FEL*, paper presented at the 11th Workshop on RF Superconductivity, 8–12 September 2003 at Travemünde/Lübeck, Germany. Online available: https://accelconf.web.cern.ch/accelconf/SRF2003/papers/weo03.pdf, last accessed 10 March 2020.

Schummer, J. *Nanotechnologie – Spiele mit Grenzen*. Frankfurt a.M.: Suhrkamp, 2009.

Schunck, H. "Bundesweite Förderung der Physik in Deutschland." In *Physik im 21. Jahrhundert: Essays zum Stand der Physik*. Edited by W. Martienssen and D. Röß, 367–403. Berlin: Springer, 2011.

Schunck, H. *Meine Jahre mit DESY*. Private Print, 2014. Online available: https://bib-pubdb1.desy.de/record/397999, last accessed 10 March 2020.

Science Board. *Minutes of the Science Board's Meeting*, 7–8 December 2009. Online available: http://www.stfc.ac.uk/files/news-from-science-board-december-2009, last accessed 1 November 2019.

Shapin, S. *The Scientific Life: A Moral History of a Late Modern Vocation*. Chicago: University of Chicago Press, 2008.

Siemann, R. "Uses of Superconductivity in Particle Accelerators." In *Lepton – Hadron Scattering, Proceedings*, 19th SLAC Summer Institute on Particle Physics (SSI 91), edited by J. Hawthorne, 21–46 (Stanford: SLAC).

Smirnov, S. "Russia – CERN Cooperation: Current Status and Perspectives." *Journal of Physics: Conference Series*, 1406, no. 1 (2019).

Stehr, N. *Knowledge Societies*. London: SAGE, 1994.

STFC. *UK Joins World Leading X-Ray Laser Facility*, News Release, 19 March 2018. Online available: https://stfc.ukri.org/news/uk-joins-world-leading-x-ray-laser-facility, last accessed 3 April 2019.

Stucke, A. "Staatliche Akteure der Wissenschaftspolitik." In *Handbuch Wissenschaftspolitik*. Edited by S. Hornbostel, A. Knie and D. Simon, 485–501. Wiesbaden: Verlag für Sozialwissenschaften, 2010.

TESLA Collaboration, *TESLA XFEL First Stage of the X-Ray Laser Laboratory: Technical Design Report Supplement*. Hamburg: DESY, 2002.

TESLA Collaboration. *TESLA. The Superconducting Electron-Positron Linear Collider with an Integrated X-Ray Laser Laboratory: Technical Design Report*. Hamburg: TESLA Collaboration, 2001.

TESLA Collaboration. *Transparencies from the TESLA Meeting*, March 1–3, 1999, Hamburg, DESY. Online available: https://flash.desy.de/sites2009/site_vuvfel/content/e403/e1644/e1418/e1420/infoboxContent2286/tesla1999-06.pdf, last accessed 10 March 2020.

TESLA Collaboration. *A Proposal to Construct and Test Prototype Superconducting RF Structures for Linear Colliders*. Hamburg: DESY, 1992.

TESLA Test Facility FEL Team. *SASE FEL at the TESLA Facility, Phase 2* (TESLA-FEL 2002-01). Hamburg: DESY, 2002.

Timmins, G. and Gower, J. "Introduction: Russia and Europe: What Kind of Partnership?" *In Russia and Europe in the Twenty-First Century: An Uneasy Partnership*. Edited by J. Gower and G. Timmins, xxi–xxvi. London: Anthem Press, 2009.

Westerlund, F. *Russian Nanotechnology R&D. Thinking Big About Small Scale Science*. Stockholm: Swedish Defence Research Agency, 2011.

Winick, H. *Synchrotron Radiation Sources: Present Capabilities and Future Directions*, SLAC-PUB-777 1, 1998.

Wissenschaftsrat. *Planungssicherheit für die Wissenschaft zur Entscheidung des Bundes über Großgeräte der Naturwissenschaftlichen Grundlagenforschung*, News Release, 5 February 2003. Online available: https://www.wissenschafts-rat.de/download/archiv/PM_0503.pdf, last accessed 10 February 2020.

Wissenschaftsrat. *Stellungnahme zu neun Großgeräten der naturwissenschaftlichen Grundlagenforschung und zur Weiterentwicklung der Investitionsplanung von Großgeräten / Statement on Nine Large-Scale Facilities for Basic Scientific Research and on the Development of Investment Planning for Large-Scale Facilities*. Köln: Wissenschaftsrat, 2003.

Wissenschaftsrat. *Systemevaluation der HGF: Stellungnahme des Wissenschaftsrates zur Hermann von Helmholtz-Gemeinschaft Deutscher Forschungszentren*. Berlin: Wissenschaftsrat, 2001.

Wissenschaftsrat. *Perspektiven für Wissenschaft und Forschung auf dem Weg zur deutschen Einheit: Zwölf Empfehlungen*. Berlin: Wissenschaftsrat, 1990.

Witte, K. *Comparison of Legal Difficulties/Advantages of ESRF and European XFEL*, presentation given at Meeting N° 2 of CTA Legal Working Group, Paris, 2012.

Womersley, J. *The STFC Science Programme 2010–15*, presentation given at the STFC Town Meeting 2009. Online available: https://www.stfc.ac.uk/files/stfc-town-meeting, last accessed 20 October 2019.

Yegorov, I. "Post-Soviet Science: Difficulties in the Transformation of the R&D Systems in Russia and Ukraine." *Research Policy* 38, no. 4 (2009): 600–609.

Yoder, J. "From Amity to Enmity: German-Russian Relations in the Post Cold War Period." *German Politics and Society* 33, no. 3 (2015): 49–69.

Zubkovskaya, A. and Michailova, S. "The Development of Russian Multinational Enterprises from the 1990s to the Present." *Organizations and Markets in Emerging Economies* 5, no. 2 (2014): 59–78.

Miscellaneous

Final Report Summary – PRE-XFEL (Preparatory Activities for the Implementation of the European X-Ray Free-Electron Laser Facility), Grant agreement ID: 211604, 1 March 2013. Online available: https://cordis.europa.eu/project/id/211604, last accessed 10 March 2020.

Website of the CREMLIN Project (Connecting Russian and European Measures for Large-Scale Research Infrastructures): https://www.cremlin.eu, last accessed 30 March 2020.

CHAPTER 6

The Other Europe of Big Science: Historical Dynamics and Contemporary Tendencies

This book investigated the founding histories of two collaborative Big Science facilities in Europe: the ESRF (European Synchrotron Radiation Facility) in Grenoble, France and the European XFEL (X-ray Free-Electron Laser) in Schenefeld, Germany. These projects constitute two cases within a much broader portfolio of Big Science projects with very different scientific purposes that were established over the course of the second half of the twentieth century and the early twenty-first century in Europe.

This book tried to advance the study of the history and politics of Big Science in mainly two regards. First, it proposed the conceptual stance of *the other Europe* to describe, study and understand the historical development of Big Science collaborations in Europe, as well as to characterise their roles within the varied European scientific and political landscapes. This approach relates to several scholarly perspectives in history and sociology that promote a more nuanced investigation of the historical developments of and in Europe, in which the history of the EEC/EU does not necessarily coincide with that of Europe. Rather, these approaches have highlighted that science, (large) technologies and infrastructures have played crucial and decisive roles in creating, shaping and projecting different kinds of Europe.[1] The conceptual approach of *the other Europe* can be summarised as follows: Although formally disentangled and institutionally independent from mainstream political integration processes and mechanisms within the EEC/EU, the history and politics of intergovernmental

K. C. Cramer, *A Political History of Big Science*,
Palgrave Studies in the History of Science and Technology,
https://doi.org/10.1007/978-3-030-50049-8_6

and collaborative Big Science projects in Europe do not stand apart. But they can be characterised as an alternative road to European integration and as crucial aspects of political and scientific activity in the recent history of Europe that mirror patterns and dynamics of bilateral and multilateral alliance-building, deal-making, controversy and compromise.

Second, this book also provided very detailed insights into the founding histories of the ESRF and the European XFEL. Importantly, it could be illustrated for both cases that national interests and strategies matter greatly to commit to collaborative Big Science efforts. This confirms what John Krige equally pinpointed namely that: "[c]ollaboration, then, and European scientific collaboration in particular, is not undertaken at the expense of self-interest; it is rather, the pursuit of one's interest by other means."[2] For instance, intergovernmental negotiations were often shaped by national political agendas and priorities, as well as broader package deals and the pursuit of so-called tit-for-tat strategies. Similarly, bargaining on a site, financial contributions and voting procedures that were recurrent issues during the founding phases of the ESRF and the European XFEL testified of usual power plays within these and similar multilateral settings. To the extent that bilateral and multilateral alliances constituted a key aspect, which has mainly driven and shaped the course of events, this also brings the unavoidable symbiosis between Big Science and politics back to geopolitical realities.

To summarise, *the other Europe* apparently exists for Big Science in Europe, and this *other Europe* mainly consists of alternative dynamics and developments beyond mainstream European politics and integration. These dynamics are thus different from those so far foregrounded by historians, sociologists or political scientists studying European history and integration, but they are important to understand the history and politics of Europe in all their complexity.

For instance, what has been characterised as a rocky relationship with the United Kingdom does not only refer to major political difficulties that surrounded initial EEC membership applications of the country in the 1960s, its subordinated role under French power plays in European politics or the withdrawal of the United Kingdom from the EU in early 2020 following a referendum in 2016. But such and similar difficult relations also run as a red threat through the founding phases of the ESRF and the European XFEL. For instance, with regard to the establishment of the ESRF, the initially proposed financial contribution from British governmental representatives was perceived as far too low by other participants, which caused severe controversies among the collaborating countries (see Sect. 4.5). In the case of the European XFEL, the sudden withdrawal of the United Kingdom from the project in late 2009 due to a national (funding) crisis put the whole effort into trouble (see Sect. 5.5). With regard to the roles of France and Germany in the early history of the ESRF, these two countries paved the way for others to join by deciding in October 1984 on initial funding and a site in Grenoble (see Sect. 4.4). This certainly was a usual bilateral demonstration of political strength and unity in Europe at a time when the two countries had also played key roles in challenging a period of deep European crisis in the early to mid-1980s (see Sect. 4.4.1). In the case of the European XFEL, Germany and Russia together pledged the necessary initial funding, after other potential partners had shown reluctance and reserved to support the project (see Sects. 5.5 and 5.6). The dominating role of Germany and Russia in the project also illustrates that the bilateral relations of the two countries constitute a major, albeit uneasy, cornerstone of the political and diplomatic landscape of Europe in the post-Cold War.

Similar dynamics and patterns can also be revealed with regard to controversy and compromise on site selection, voting procedures and financial shares (see Sects. 2.1.3, 4.3 and 5.5). With regard to the site selection process during the founding phase of the ESRF, the partner countries of the project clearly proposed and/or accepted those sites either that were located on their own national territory or that they regarded as the most beneficial ones. For instance, France proposed Strasbourg and later Grenoble, whereas Italy went for Trieste, and Denmark for Risø. Germany supported the candidature of Strasbourg because it is located very close to the French-German border. The United Kingdom proposed Daresbury but also agreed on Grenoble because this meant a co-location with the ILL, in which the United Kingdom participates as one out of three members. The Nordic countries Sweden and Norway supported the Danish application of Risø obviously because it was located close to their own countries. With regard to the site selection for the European XFEL, there seemingly was no extensive site selection process, but there was not much doubt that the free-electron laser would be closely connected to the TESLA linear collider at DESY in Hamburg. It can only be speculated to what extent the absence of any site selection process impacted the attitude of many potential partner countries that were reluctant to join the project (see Sect. 5.5). Why should they substantially fund a project that would not be located on their own territory and to which their scientists would clearly have access depending on the successful peer-review process of their applications?

With regard to the increasingly important role of Russia in European Big Science collaborations, the end of the Cold War certainly had a role to play. Russian membership in the European XFEL project or a similar effort, as well as a substantial financial contribution of a similar scale, would certainly have been unthinkable a few decades earlier, for instance, in the founding phase of the ESRF during the late Cold War. The large Russian contribution to the European XFEL project, which became financially dominated by Germany and Russia, is not only a visible testament of how the end of the Cold War translated into re-framed alliance-building on the European continent. But it also illustrated how the crucial political importance of Russia for post-Cold War Europe resonated in scientific collaboration, making Germany and Russia close partners in both politics and Big Science. Moreover, there is much to suggest that Russia participates in the European XFEL project, as well as in similar other projects, such as FAIR or the ESRF, because all of these projects were located in Western Europe. It seems reasonable to argue that for Russia, investment in (Western) European Big Science projects carried strong territorial and/or spatial implications, namely, that political commitment to these Big Science facilities also constituted a projection of national power and security interests in the post-Cold War. Moreover, it was (and remains) a strategy for the country to ask for a return of investment on national Big Science projects planned to be built on Russian territory in the context of the Russian Megascience Initiative (see Sect. 5.6). But while Russia's full membership in Big Science collaborations in Europe only during the very recent decades became an accepted new reality of post-Cold War Europe, scientific collaboration between East and West was not at all absent during the Cold War. In contrast, major research institutes such as CERN or DESY maintained a lively exchange of scientists and ideas with the Soviet Union throughout the Cold War. The contours of these efforts were, however, also bound to the political contexts of their time, which partly explains why East/West collaboration during the Cold War remained to be based on individual initiatives and smaller projects, and could not escalate to the intergovernmental level.

Certainly, as has been illustrated throughout this book, politics are key. But exclusive emphasis on political aspects misses to take into account other crucial concerns: First, personal ties played fundamental and decisive roles in the founding phases of the ESRF and the European XFEL. From personal initiatives of scientists such as Heinz Maier-Leibnitz or Björn Wiik, the influential voice of personalities such as the politician Louis Mermaz, to the role of administrators in the ministries and governments such as state secretary Frieder Meyer-Krahmer or general director Hermann Schunck—all of whom played decisive roles in making the ESRF or the European XFEL a reality. Second, the use of particle accelerators not for particle physics research but for research with synchrotron radiation became an increasingly demanded technology and experimental resources throughout the last decades. This development did not only alter the scientific landscapes in Europe with several new, dedicated synchrotron radiation sources that were built from the late 1980s and early 1990s onwards. But is also resonates a new line of political reasoning in the post-Cold War that preferred application-oriented experimental research at, for instance, synchrotron radiation sources or free-electron lasers, over fundamental investigations in particle physics because the former connects more closely to solving the grand challenges of today, and to an ever-more strategically oriented science policy regime (see Sects. 5.4 and 2.2).

With regard to the role of the EEC/EU, it can be argued that the EEC lacked any competences to shape the creation of the ESRF in the late 1970s and mid-1980s. Based on the findings in Chap. 4, it can, however, be argued that the French-German partnership filled this gap (as it had often done in the past). The two countries acted as a catalyser, as science administrator Pierre Papon called it, for the construction of a European space for research.[3] Since the early 2000s, the EU started to implement several measures, strategies and tools that unfolded in what can be characterised as Research Infrastructures (RIs) policy. However, apart from partly funding the preparatory phase of the European XFEL project, the EU did not have a role to play in the project's early history. But political interest in the issue of Research Infrastructures (which partly overlaps with that of Big Science) probably links to changing roles of science, technology and research in the post-Cold War period, namely their more strategic role for and within economy and society, and their potential usefulness for creating commercial applications and for solving grand societal challenges (see Chaps. 1 and 2).

The discussed aspects of the founding phases of the ESRF and the European XFEL are, however, not only tales of the recent past of Europe. But current developments equally testify how Big Science is unavoidably linked to European and international politics and diplomacy. This is not only because politics heavily matter in collaborative Big Science but also, as mentioned above, because governments and administrators expect these projects to nowadays play decisive roles in economy and society.

For instance, in the mid-2010s, Hungary encountered severe difficulties in providing its annual membership fees to the ESRF. This situation was not necessarily due to a lack of budget but rather to the political situation in Hungary at that time. The country experienced a strong anti-European political climate and authoritarian political regime after Hungarian Prime Minister Victor Orban had climbed to power in 2010. Public funding, so it seems, was not to support a collaborative European effort.[4] Another example is the recent withdrawal of Russia from its application as an associate country of CERN in order to negotiate a new agreement that should give the country "a special status for participation in the experiments."[5] The agreement, moreover, "will include CERN's involvement in construction of mega science facilities in Russia."[6] Similarly, when Russia became a full member of the ESRF project in 2014, contributing 6 per cent of the operation costs, this led to a large re-allocation process resulting in lower contributions by other European members of the project. Both examples further support one of the main findings and core messages of this book, namely, that national politics and policies matter significantly for an understanding of the patterns and dynamics of collaborative Big Science projects in Europe. After the withdrawal of the United Kingdom from the European XFEL project in 2009, the British flag was nevertheless displayed among the flags of the other member countries during the European XFEL's opening ceremony in September 2017.[7] This happened before the country officially re-joined the project in 2018. Despite the anecdotal character, this example is a strong sign that although politicians and governmental representatives could not agree on formal membership and official contribution, scientists apparently kept on collaborating.

With regard to the very recent *Brexit*, the exit of the United Kingdom from the European Union in early 2020, it is certainly too early to assess the full consequences for scientific collaboration in Europe. But it should be borne in mind that previous scholarly research illustrated that the United Kingdom has always been both a crucial and an uneasy partner in Big Science collaborations in Europe.[8] In light of the recent COVID-19 pandemic, it can only be speculated how and to what extent Big Science facilities will have a role to play. This particularly matters in two regards.

First, single-sited Big Science facilities, such as synchrotron radiation sources or free-electron lasers, are geographically bound because they host large scientific instruments that are unable to move. Yet, in turn, they require people to re-locate for the purpose of conducting experiments. But the need to collaborate *on site* and to travel in order to arrange and manage experimental settings seemingly runs counter the call for *physical distancing*. Second, it is questionable how and to what extent COVID-19-related research at Big Science facilities may (or may not) match the high political expectations that were put on these and similar projects in recent decades. Policymakers and governments claimed in recent years that these projects should considerably contribute to the solving of urgent societal challenges such as climate change, health or energy security that sociologist Olof Hallonsten characterised as a "political hype"[9] rather than a sustained assessment of the real capabilities and limits of these and similar scientific projects.

Notes

1. See, for example, T. Misa and J. Schot, "Inventing Europe: Technology and the Hidden Integration of Europe." *History and Technology* 21, no. 1 (2005); H. Trischler and H. Weinberger, "Engineering Europe: Big Technologies and Military Systems in the Making of 20th Century Europe." *History and Technology* 21, no. 1 (2005); F. Schipper and J. Schot "Infrastructural Europeanism, or the Project of Building Europe on Infrastructures: An Introduction." *History and Technology* 27, no. 3 (2011); K. Patel, "Provincialising European Union: Co-Operation and Integration in Europe in a Historical Perspective." *Contemporary European History* 22, no. 04 (2013), 650.
2. J. Krige, "The Politics of European Scientific Collaboration." In *Companion to Science in the Twentieth Century,* eds. J. Krige and D. Pestre (London: Routledge, 2003), 900.
3. P. Papon, "L'Espace Européen de la Recherche (1960–1985): Entre Science et Politique." In *La Construction d'un Espace Scientifique Commun? La France, la RFA et l'Europe après le "Choc du Spoutnik"*, eds. C. Defrance and U. Pfeil (Bruxelles, New York: P.I.E. Peter Lang, 2012), 42.
4. See, Petition of the Hungarian Synchrotron Committee at change.org: *Stop Hungary's withdrawal from the European Synchrotron Radiation Facility.*

5. Joint Institute for Nuclear Research, "Russia and CERN are Working Out a New Format of Cooperation," News Release (14 March 2018).
6. TASS, "New Agreement with CERN to be Signed in 2018 – Russia's Education Ministry," News Release, 10 March 2018.
7. J. Amos, "XFEL: Brilliant X-Ray Laser Comes Online." *BBC News*, 1 September 2017.
8. See, for example, K. C. Cramer, "The Role of European Big Science in the (Geo)Political Challenges of the Twentieth and Twenty-First Centuries." In *Big Science and Research Infrastructures in Europe*, eds. K. C. Cramer and O. Hallonsten (Cheltenham: Edward Elgar, 2020).
9. O. Hallonsten, "Research Infrastructures in Europe: The Hype and the Field." *European Review* 28, no. 4 (2020).

Bibliography

Amos, J. "XFEL: Brilliant X-Ray Laser Comes Online." *BBC News*, 1 September 2017. Online available: https://www.bbc.com/news/science-environment-41117442, last accessed 12 December 2019.

Cramer, K. C. "The Role of European Big Science in the (Geo)Political Challenges of the Twentieth and Twenty-First Centuries." In *Big Science and Research Infrastructures in Europe*, eds. K. C. Cramer and O. Hallonsten, 56–75. Cheltenham: Edward Elgar, 2020.

Hallonsten, O. "Research Infrastructures in Europe: The Hype and the Field." *European Review* 28, no. 4 (2020): 617–635.

Joint Institute for Nuclear Research. "Russia and CERN are Working Out a New Format of Cooperation," News Release, 14 March 2018. Online available: http://www.jinr.ru/posts/russia-and-cern-are-working-out-a-new-format-of-cooperation, last accessed 20 March 2020.

Krige, J. "The Politics of European Scientific Collaboration." In *Companion to Science in the Twentieth Century*, edited by J. Krige and D. Pestre, 897–918. London: Routledge, 2003.

Misa T., and Schot, J. "Inventing Europe: Technology and the Hidden Integration of Europe." *History and Technology* 21, no. 1 (2005): 1–19.

Papon, P. "L'Espace Européen de la Recherche (1960–1985): Entre Science et Politique." In *La Construction d'un Espace Scientifique Commun? La France, la RFA et l'Europe après le "Choc du Spoutnik"*, edited by C. Defrance and U. Pfeil, 37–54. Bruxelles, New York: P.I.E. Peter Lang, 2012.

Patel, K. K. "Provincialising European Union: Co-Operation and Integration in Europe in a Historical Perspective." *Contemporary European History* 22, no. 4 (2013): 649–673.

Schipper, F., and Schot, J. "Infrastructural Europeanism, or the Project of Building Europe on Infrastructures: An Introduction." *History and Technology* 27, no. 3 (2011): 245–264.

Trischler, H., and Weinberger, H. "Engineering Europe: Big Technologies and Military Systems in the Making of 20th Century Europe." *History and Technology* 21, no. 1 (2005): 49–83.

TASS. "New Agreement with CERN to be Signed in 2018 – Russia's Education Ministry." News Release, 10 March 2018. Online available: http://tass.com/science/993418, last accessed 12 Mai 2020.

Miscellaneous

Petition of the Hungarian Synchrotron Committee at change.org: *Stop Hungary's withdrawal from the European Synchrotron Radiation Facility*. Online available: https://www.change.org/p/government-of-hungary-hungarian-academy-of-sciences-stop-hungary-s-withdrawal-from-the-european-synchrotron-radiation-facility, last accessed 30 March 2020.

Bibliography

Chapter 1

Bush, V. *Science: The Endless Frontier.* Washington: Government Printing Office, 1945.

Capshew, J. and Rader, K. "Big Science: Price to the Present." *Osiris* 7, Science after '40 (1992a): 2–25.

Castells, M. and Cardoso, G., eds. *The Network Society: From Knowledge to Policy.* Washington DC: Johns Hopkins Center for Transatlantic Relations, 2005a.

Cramer, K. C. "The Role of European Big Science in the (Geo)Political Challenges of the Twentieth and Twenty-First Centuries." In *Big Science and Research Infrastructures in Europe*, edited by K. C. Cramer and O. Hallonsten. Cheltenham: Edward Elgar, forthcoming 2020a.

Cramer, K. C. "Lightening Europe: Establishing the European Synchrotron Radiation Facility (ESRF)." *History and Technology* 33, no. 4 (2017a): 396–427.

Cramer, K. C. and Hallonsten, O. "Big Science and Research Infrastructures in Europe: Conclusions and Outlook." In *Big Science and Research Infrastructures in Europe*, edited by K. C. Cramer and O. Hallonsten. Cheltenham: Edward Elgar, forthcoming 2020.

Cramer, K. C., Hallonsten, O., Bolliger, I., and Griffiths, A. "Big Science and Research Infrastructures in Europe: History and Current Trends." In *Big Science and Research Infrastructures in Europe*, edited by K. C. Cramer and O. Hallonsten. Cheltenham: Edward Elgar, forthcoming 2020.

Crease, R. "The National Synchrotron Light Source: Part I: Bright Idea." *Physics in Perspective* 9 (2007): 1–30.

K. C. Cramer, *A Political History of Big Science*,
Palgrave Studies in the History of Science and Technology,
https://doi.org/10.1007/978-3-030-50049-8

Crease, R. "The National Synchrotron Light Source: Part II: The Bakeout." *Physics in Perspective* 10 (2008): 1–31.

Crease, R. and Westfall, C. "The New Big Science." *Physics Today* 69, no. 5 (2016): 30–36.

Defrance, C. "France-Allemagne: Une Coopération Scientifique 'Privilégiée' en Europe, de l'Immédiat Après-Guerre au Milieu des Années 1980?" In La Guerre Froide et l'Internationalisation des Sciences: Acteurs, Réseaux et Institutions. Edited by C. Defrance and A. Kwaschik, 169–186. Paris: CNRS, 2016a.

Doing, P. *Velvet Revolution at the Synchrotron: Biology, Physics, and Change in Science*. Cambridge, MA: MIT Press, 2009.

Duşa, A., Nelle, D., Stock, G. and Wagner G. G., eds., *Facing the Future. European Research Infrastructures for the Humanities and Social Sciences*. Berlin: Scivero Verlag, 2014.

Eggleton, D. *Examining the Relationship Between Leadership and Megascience Projects*. Doctoral Thesis: University of Sussex, 2017.

Elzinga, A. "Features of the Current Science Policy Regime: Viewed in Historical Perspective." *Science and Public Policy* 39, no. 4 (2012a): 416–28.

European Commission. *Developing World-class Research Infrastructures for the European Research Area (ERA): Report of the ERA Expert Group*. Luxembourg: Office for Official Publications of the European Communities, 2008a.

Forman, P. "Behind Quantum Electronics: National Security as Basis for Physical Research in the United States, 1940–1960." *Historical Studies in the Physical and Biological Sciences* 18, no. 1 (1987): 149–229.

Franssen, T. "Research Infrastructure Funding as a Tool for Science Governance in the Humanities: A Country Case Study of the Netherlands." In *Big Science and Research Infrastructures in Europe*, edited by K. Cramer and O. Hallonsten. Cheltenham: Edward Elgar, forthcoming 2020.

Galison, P. "The Many Faces of Big Science." In *Big Science: The Growth of Large-Scale Research*, edited by P. Galison and B. Hevly, 1–17. Stanford: Stanford University Press, 1992.

Galison, P. and Hevly, B., eds., *Big Science: The Growth of Large-Scale Research*. Stanford: Stanford University Press, 1992a.

Galison, P. *Image and Logic: A Material Culture of Microphysics*. Chicago: University of Chicago Press, 1997.

Hallonsten, O. "Continuity and Change in the Politics of European Scientific Collaboration." *Journal of Contemporary European Research* 8, no. 3 (2012a): 300–319.

Hallonsten, O. "Research Infrastructures in Europe: The Hype and the Field." *European Review* (2020a).

Hallonsten, O. "The Parasites." *Historical Studies in the Natural Sciences* 45, no. 2 (2015a): 217–272.

Hallonsten, O. "The Politics of European Collaboration in Big Science." In *The Global Politics of Science and Technology – Vol. 2*. Edited by M. Mayer, M. Carpes and R. Knoblich, 31–46. Berlin, Heidelberg: Springer, 2014a.

Hallonsten, O. *Big Science Transformed: Science, Politics and Organization in Europe and the United States.* Cham: Palgrave Macmillan, 2016a.

Hallonsten, O. *Small Science on Big Machines: Politics and Practices of Synchrotron Radiation Laboratories.* Lund: Research Policy Institute, 2009a.

Hallonsten, O., Eriksson, H., and Collsiöö, A. "The Role of Research Infrastructures in Innovation Systems: The Case of Swedish Participation in the Halden Reactor Project (HRP)." In *Big Science and Research Infrastructures in Europe*, edited by K. Cramer and O. Hallonsten. Cheltenham: Edward Elgar, forthcoming 2020.

Heinze, T., Hallonsten, O., and Heinecke, S. "Turning the Ship: The Transformation of DESY, 1993–2009." *Physics in Perspective* 19, no. 4 (2017a): 424–451.

Hevly, B. "Reflections on Big Science and Big History." In *Big Science: The Growth of Large-Scale Research*, edited by P. Galison and B. Hevly, 355–364. Stanford: Stanford University Press, 1992.

Hoddeson, L., Kolb, A. and Westfall, C. *Fermilab: Physics, the Frontier, and Megascience.* Chicago: University of Chicago Press, 2008a.

Jacob, M. and Hallonsten, O. "The Persistence of Big Science and Megascience in Research and Innovation Policy." *Science and Public Policy* 39, no. 4 (2012): 411–415.

Joint Institute for Nuclear Research. "Russia and CERN are Working Out a New Format of Cooperation," News Release, 14 March 2018. Online available: http://www.jinr.ru/posts/russia-and-cern-are-working-out-a-new-format-of-cooperation, last accessed 20 March 2020.

Judt, T. *Postwar: A History of Europe since 1945.* New York et al.: Penguin Press, 2005.

Kaiserfeld, T. and O'Dell, T., eds. *Legitimizing ESS: Big Science as a Collaboration Across Boundaries.* Lund: Nordic Academic Press, 2013.

Kehrt, C. *Mit Molekülen spielen: Wissenschaftskulturen der Nanotechnologie zwischen Politik und Medien* (Bielefeld: transcript Verlag, 2016a).

Kevles, D. "Cold War and Hot Physics: Science, Security, and the American State, 1945–56." *Historical Studies in the Physical and Biological Sciences* 20, no. 2 (1990): 239–264.

Koopmann, M. and Schild, J. "Eine neue Ära? Deutsch-Französische Beziehungen nach dem Ende des Kalten Krieges." In *Neue Wege in ein Neues Europa: Die Deutsch-Französischen Beziehungen nach dem Ende des Kalten Krieges*, edited by M. Koopmann, J. Schild and H. Stark, 199–210. Baden-Baden: Nomos, 2013.

Krige, J., ed. *Choosing Big Technologies.* Chur: Harwood Academic Publishers, 1993a.

Krige, J. "Preface." In *Choosing Big Technologies*, edited by J. Krige, vii–xii. Chur: Harwood Academic Publishers, 1993b.

Krige, J. "The Politics of European Scientific Collaboration." In *Companion to Science in the Twentieth Century*, edited by J. Krige and D. Pestre, 897–918. London: Routledge, 2003a.

Krige, J. *American Hegemony and the Postwar Reconstruction of Science in Europe*. Cambridge, MA: MIT Press, 2006a.

Krige, J. and Guzzetti, L., eds. *History of European Scientific and Technological Cooperation*. Luxembourg: European Communities, 1997.

Krotz, U. and Schild, J. *Shaping Europe: France, Germany, and Embedded Bilateralism from the Elysée Treaty to Twenty-First Century Politics*. Oxford: Oxford University Press, 2013a.

Kühnhardt, L. *European Union – The Second Founding: The Changing Rationale of European Integration*. Baden-Baden: Nomos, 2008.

Lengwiler, M. "Kontinuitäten und Umbrüche in der Deutschen Wissenschaftspolitik des 20. Jahrhunderts." In *Handbuch Wissenschaftspolitik*, edited by S. Hornbostel, A. Knie and D. Simon, 6–18. Wiesbaden: Verlag für Sozialwissenschaften, 2010a.

Lohrmann, E. and Söding, P. *Von schnellen Teilchen und hellem Licht: 50 Jahre Deutsches Elektronen-Synchrotron DESY*. Weinheim: Wiley, 2009a.

May, A. *Britain and Europe since 1945*. Hoboken: Taylor and Francis, 2014.

Middlemas, K. *Orchestrating Europe: The Informal Politics of the European Union 1973–1995*. London: Fontana Press, 1995a.

Misa, T. and Schot, J. "Inventing Europe: Technology and the Hidden Integration of Europe." *History and Technology* 21, no. 1 (2005a): 1–19.

Nickelsen, K. and Krämer, F. "Introduction: Cooperation and Competition in the Sciences," *NTM Zeitschrift für Geschichte der Wissenschaften, Technik und Medizin* 24 (2016): 119–123.

Papon, P. "European Scientific Cooperation and Research Infrastructures: Past Tendencies and Future Prospects." *Minerva* 42, no. 1 (2004a): 61–76.

Papon, P. "Intergovernmental Cooperation in the Making of European Research." In *European Science and Technology Policy: Towards Integration or Fragmentation?* edited by H. Delanghe, U. Muldur, and L. Soete, 24–43. Cheltenham: Edward Elgar, 2009a.

Papon, P. "L'Espace Européen de la Recherche (1960–1985): Entre Science et Politique." In *La Construction d'un Espace Scientifique Commun? La France, la RFA et l'Europe après le "Choc du Spoutnik"*. Edited by C. Defrance and U. Pfeil, 37–54. Bruxelles, New York: P.I.E. Peter Lang, 2012a.

Papon, P. *L'Europe de la Science et de la Technologie*. Grenoble: Press Universitaires de Grenoble (PUG), 2001a.

Patel, K. "Rockefeller Foundation, Kalter Krieg und Amerikanisierung." In *American Foundations and the Coproduction of World Order in the Twentieth Century*, edited by H. Rausch and J. Krige, 173–185. Göttingen: Vandenhoeck & Ruprecht, 2012.

Pestre, D. "Science, Political Power and the State." In *Companion to Science in the Twentieth Century.* Edited by J. Krige and D. Pestre, 61–75. London: Routledge, 2003.

Pollock, H. "The Discovery of Synchrotron Radiation." *American Journal of Physics* 51, no. 3 (1983a): 278–280.

Praderie, F. "Big Science: Why? Where? and How?" *Memorie della Società Astronomia Italiana* 67, no. 4 (1996): 889–899.

Price, D. K. *Government and Science: Their Dynamic Relation in American Democracy.* New York: NYU Press, 1954

Price, Derek J. de Solla. *Little Science, Big Science.* New York: Columbia University Press, 1963.

Ravetz, J. *Scientific Knowledge and its Social Problems.* New York: Oxford University Press, 1971.

Rekers, J. and Sandell, K., eds. *New Big Science in Focus.* Lund: Lund University Press, 2016.

Riordan, M., Hoddeson, L. and Kolb, A. *Tunnel Visions: The Rise and Fall of the Superconducting Super Collider.* Chicago: University of Chicago Press, 2015a.

Ritter, G. *Großforschung und Staat in Deutschland: Ein historischer Überblick.* München: Beck, 1992a.

Schipper, F. and Schot, J. "Infrastructural Europeanism, or the Project of Building Europe on Infrastructures: An Introduction." *History and Technology* 27, no. 3 (2011a): 245–264.

Schmied, H. "The European Synchrotron Radiation Story – Phase II." *Synchrotron Radiation News* 3, no. 6 (1990a): 22–26.

Schmied, H. "The European Synchrotron Radiation Story." *Synchrotron Radiation News* 3, no. 1 (1990b): 18–22.

Seidel, R. W. "A Home for Big Science: The Atomic Energy Commission's Laboratory System." *Historical Studies in the Physical and Biological Sciences* 16, no. 1 (1986): 135–175.

Seidel, R. W. "The National Laboratories of the Atomic Energy Commission in the early Cold War." *Historical Studies in the Physical and Biological Sciences* 32, no. 1 (2001): 145–162.

Shapin, S. *The Scientific Life: A Moral History of a Late Modern Vocation.* Chicago: University of Chicago Press, 2008a.

Simoulin, V. *Sociologie d'un Grand Équipement Scientifique: Le Premier Synchrotron de Troisième Génération.* Lyon: ENS Éditions, 2012.

Stehr, N. *Knowledge Societies.* London: SAGE, 1994a.

Szöllösi-Janze, M. and Trischler, H. *Großforschung in Deutschland.* Frankfurt, New York: Campus Verlag, 1990.

Timmins, G. and Gower, J. "Introduction: Russia and Europe: What Kind of Partnership?" *In Russia and Europe in the Twenty-First Century: An Uneasy Partnership.* Edited by J. Gower and G. Timmins, xxi-xxvi. London: Anthem Press, 2009a.

Trischler, H. and Weinberger, H. "Engineering Europe: Big Technologies and Military Systems in the Making of 20th Century Europe." *History and Technology* 21, no. 1 (2005a): 49–83.

Ulnicane, I. "Ever-changing Big Science and Research Infrastructures: Evolving EU Policy." In *Big Science and Research Infrastructures in Europe*, edited by K. Cramer and O. Hallonsten. Cheltenham: Edward Elgar, forthcoming 2020.

Venter, C. and Cohen, D. "The Century of Biology." *New Perspectives Quarterly* 21, no. 4 (2004): 73–77.

Vermeulen, N. *Supersizing Science: On Building Large-Scale Research Projects in Biology*. Gardners Books, 2010.

Weinberg, A. "Impact of Large-Scale Science on the United States." *Science* 134, no. 3473 (1961): 161–164.

Weinberg, A. *Reflections on Big Science*. Cambridge, Mass.: MIT Press, 1967.

Westfall, C. "Institutional Persistence and the Material Transformation of the US National Labs: The Curious Story of the Advent of the Advanced Photon Source." *Science and Public Policy* 39, no. 4 (2012): 439–449.

Westfall, C. "Rethinking Big Science: Modest, Mezzo, Grand Science and the Development of the Bevalac, 1971–1993." *ISIS: Journal of the History of Science in Society* 94, no. 1 (2003): 30.

Westfall, C. and Hoddeson, L. "Thinking Small in Big Science: The Founding of Fermilab, 1960–1972." *Technology and Culture* 37, no. 3 (1996): 457–492

Westfall, C. and Krige, J. "The Path of Post-War Physics." In *The Particle Century*. Edited by G. Fraser, 1–11. Bristol, Philadelphia: Institute of Physics Pub., 1998.

Westwick, P. *The National Labs: Science in an American System, 1947–1974*. Cambridge, MA: Harvard University Press, 2003.

Winick, H., Doniach, S. "An Overview of Synchrotron Radiation Research." In *Synchrotron Radiation Research*, edited by H. Winick and S. Doniach, 1–10. Boston, MA: Springer, 1980a.

Zilsel, P. "The Mass Production of Knowledge." *Bulletin of the Atomic Scientists* 20, no. 4 (1964): 28–29.

Chapter 2

Ash, M. "Wissenschaft und Politik als Ressourcen füreinander." In *Wissenschaften und Wissenschaftspolitik: Bestandsaufnahmen zu Formationen, Brüchen und Kontinuitäten im Deutschland des 20. Jahrhunderts*, edited by R. vom Bruch and B. Kaderas, 32–51. Stuttgart: Franz Steiner, 2002.

Ash, M. "Wissenschaft und Politik: Eine Beziehungsgeschichte im 20. Jahrhundert." *Archiv für Sozialgeschichte* 50 (2010): 11–46.

Badenoch, A. and Fickers, A. *Materializing Europe: Transnational Infrastructures and the Project of Europe*. New York, NY: Palgrave Macmillan, 2010.

Banchoff, T. "Political Dynamics of the ERA." In *Changing Governance of Research and Technology Policy: The European Research Area*, edited by J. Edler, S. Kuhlmann, and M. Behrens, 81–97. Cheltenham: Edward Elgar, 2003.

Bolliger, I. and Griffiths, A. "The Introduction of ESFRI and the Rise of National Research Infrastructure Roadmaps in Europe." In *Big Science and Research Infrastructures in Europe*, edited by K. Cramer and O. Hallonsten. Cheltenham: Edward Elgar, forthcoming 2020.

Capshew, J. and Rader, K. "Big Science: Price to the Present." *Osiris* 7, Science after '40 (1992b): 2–25.

Castelnovo, P., Florio, M., Forte, S., Rossi, L., and Sirtori, E. The Economic Impact of Technological Procurement for Large-Scale Research Infrastructures: Evidence from the Large Hadron Collider at CERN. *Research Policy* 47, no. 9 (2018): 1853–1867.

Cramer, K. "The Role of European Big Science in the (Geo)Political Challenges of the Twentieth and Twenty-First Centuries." In *Big Science and Research Infrastructures in Europe*, edited by K. Cramer and O. Hallonsten. Cheltenham: Edward Elgar, forthcoming 2020b.

de Elera, A. "The European Research Area: On the Way Towards a European Scientific Community?" *European Law* 12, no. 5 (2006): 559–574.

Defrance, C. "France-Allemagne: Une Coopération Scientifique 'Privilégiée' en Europe, de l'Immédiat Après-Guerre au Milieu des Années 1980?" In *La Guerre Froide et l'Internationalisation des Sciences: Acteurs, Réseaux et Institutions*, edited by C. Defrance and A. Kwaschik, 169–186. Paris: CNRS, 2016b.

Delanghe, H., Muldur, U., and Soete L. "Conclusion and Perspectives." In *European Science and Technology Policy: Towards Integration or Fragmentation?* edited by H. Delanghe, U. Muldur, and L. Soete, 353–56. Cheltenham: Edward Elgar, 2009.

ESF, ed. *ESF EUROCORES Programme "Inventing Europe – Technology and the Making of Europe, 1850 to the Present": Final Report.* Strasbourg: European Science Foundation, 2011.

European Commission. *Developing World-class Research Infrastructures for the European Research area (ERA): Report of the ERA Expert Group* (Luxembourg: Office for Official Publications of the European Communities, 2008b.

European Commission. *Growth, Competitiveness, Employment. The Challenges and Ways Forward into the 21st Century*, White paper. Luxemburg: Office for Official Publications of the European Commission, 1993.

European Commission. *Communication from the Commission to the Council, the European Parliament, the Economic and Social Committee and the Committee of the Regions – Towards a European Research Area*. Brussels, 2000.

Fazekas, A. "Big Science in a Small Country." *Science*, 20 January 2006. Online available at: https://www.sciencemag.org/careers/2006/01/big-science-small-country (last accessed 10 December 2019.

Goldsmith, M. "Variable Geometry, Multilevel Governance: European Integration and Subnational Government in the New Millennium." In *The Politics of Europeanization*, edited by K. Featherstone and C. Radaelli, 112–133. Oxford University Press, 2003.

Guzzetti, L. *A Brief History of European Union Research Policy.* Luxembourg: European Communities, 1995a.

Hacking, I. "The Self-Vindication of the Laboratory Sciences." In *Science as Practice and Culture*, edited by A. Pickering, 29–64. Chicago: University of Chicago Press, 1992.

Hallonsten, O. "Is There an 'Iron Law' of Big Science?" In *Big Science and Research Infrastructures in Europe*, edited by K. Cramer and O. Hallonsten. Cheltenham: Edward Elgar, forthcoming 2020c.

Hallonsten, O. "Research Infrastructures in Europe: The Hype and the Field." *European Review* (2020d), 1–19.

Hallonsten, O. *Big Science Transformed: Science, Politics and Organization in Europe and the United States.* Cham: Palgrave Macmillan, 2016b.

Hallonsten, O. "The Politics of European Collaboration in Big Science." In *The Global Politics of Science and Technology*, Vol. 2, edited by M. Mayer, M. Carpes, and R. Knoblich, 31–46. Berlin, Heidelberg: Springer, 2014b.

Hallonsten, O. "Myths and Realities of the ESS project: A Systematic Scrutiny of Readily Accepted 'Truths'." In *Legitimizing ESS: Big Science as a Collaboration Across Boundaries*, edited by T. Kaiserfeld, and T. O'Dell, 43–66. Lund: Nordic Academic Press, 2013.

Hallonsten, O. "Contextualising the European Spallation Source: What We Can Learn from the History, Politics, and Sociology of Big Science." In *In Pursuit of a Promise: Perspectives on the Political Process to Establish the European Spallation Source (ESS) in Lund, Sweden.* Edited by O. Hallonsten, 81–107. Lund: Arkiv Academic Press, 2012c.

Hallonsten, O. "Continuity and Change in the Politics of European Scientific Collaboration." *Journal of Contemporary European Research* 8, no. 3 (2012d): 300–319.

Heymann, M. and Martin-Nielsen, J. "Introduction: Perspectives on Cold War Science in Small European States." *Centaurus* 55 (2013): 221–242.

Hoddeson, L., Kolb, A. and Westfall, C. *Fermilab: Physics, the Frontier, and Megascience.* Chicago: University of Chicago Press, 2008b.

Jasanoff, S. "Ordering Knowledge, Ordering Society." In *States of Knowledge: The Co-production of Science and the Social Order*, edited by S. Jasanoff, 13–45. London: Routledge, 2006.

Krige, J. "Embedding the National in the Global." In *Science and Technology in the Global Cold War*, edited by N. Oreskes and J. Krige, 227–250. Cambridge, MA: MIT Press, 2014.

Krige, J. "The Politics of European Scientific Collaboration." In *Companion to Science in the Twentieth Century*, edited by J. Krige and D. Pestre, 897–918. London: Routledge, 2003b.

Krige, J. "Historical Synthesis." In *History of European Scientific and Technological Cooperation*, edited by J. Krige and L. Guzzetti, 439–444. Luxembourg: European Communities, 1997.

Krige, J. "Preface." In *Choosing Big Technologies*, edited by J. Krige, vii-xii. Chur: Harwood Academic Publishers, 1993.

Krige, J. and D. Pestre (eds.). *Companion to Science in the Twentieth Century* (London: Routledge, 2003).

Li-Hua, R. "Definitions of Technology." In *A Companion to the Philosophy of Technology*, edited by J. Friis, S. Pedersen, and V. Hendricks, 18–22. Chichester: Wiley-Blackwell, 2009.

Luukkonen, T. "European Research Area: An Evolving Policy Agenda." In *Towards European Science. Dynamics and Policy of an Evolving European Research Space*, edited by L. Wedlin and M. Nedeva, 37–60. Cheltenham: Edgward Elgar, 2015.

McCray, W. "'Globalization with Hardware': ITER's Fusion of Technology, Policy, and Politics." *History and Technology* 26, no. 4 (2010a): 283–312.

Misa T., and Schot, J. "Inventing Europe: Technology and the Hidden Integration of Europe." *History and Technology* 21, no. 1 (2005b): 1–19.

Moskovko, M. "Intensified Role of the EU? European Research Infrastructure Consortium (ERIC) as a Legal Framework for Contemporary Multinational Research Collaboration. In *Big Science and Research Infrastructures in Europe*, edited by K. Cramer and O. Hallonsten. Cheltenham: Edward Elgar, forthcoming 2020.

Moskovko, M., Astvaldsson, A., and Hallonsten, O. "Who is ERIC? The Politics and Jurisprudence of a New Governance Tool for Collaborative European Research Infrastructures." *Journal of Contemporary European Research* 15, no. 3 (2019): 249–268.

Nedeva, N. "Between the Global and the National: Organising European Science." *Research Policy* 42, no. 1 (2013): 220–230.

OECD. *The Research System: Comparative Survey of the Organization and Financing of Fundamental Research*, Volume I: France, Germany, United Kingdom; Volume II: Belgium, Netherlands, Sweden, Switzerland; Volume III: Canada, United States, General Conclusions. Paris: OECD Publishing, 1972–1974.

OECD. *Gaps in Technology. Analytical Report: Comparison between Member Countries in Education, Research Development, Technological Innovation, International Economic Exchange*. Paris: OECD Publishing, 1970.

OECD. *The Measurement of Scientific and Technical Activities: Proposed Standard Practice for Surveys of Research and Experimental Development.* Paris: OECD Publishing, 1962.

Papon, P. "L'Espace Européen de la Recherche (1960–1985): Entre Science et Politique." In *La Construction d'un Espace Scientifique Commun? La France, la RFA et l'Europe après le "Choc du Spoutnik"*, edited by C. Defrance and U. Pfeil, 37–54. Bruxelles, New York: P.I.E. Peter Lang, 2012b.

Papon, P. "Intergovernmental Cooperation in the Making of European Research." In *European Science and Technology Policy: Towards Integration or Fragmentation?* edited by H. Delanghe, U. Muldur, and L. Soete, 24–43. Cheltenham: Edward Elgar, 2009b.

Papon, P. "European Scientific Cooperation and Research Infrastructures: Past Tendencies and Future Prospects." *Minerva* 42, no. 1 (2004b): 61–76.

Patel, K. K. "Provincialising European Union: Co-Operation and Integration in Europe in a Historical Perspective." *Contemporary European History* 22, no. 4 (2013a): 649–673.

Pero, H. "Research Infrastructures of Pan-European Interest: The EU and Global Issues." *Nuclear Instruments and Methods in Physics Research Section A* 626–627 (2011): 69–71.

Peterson, J. "Eureka: A Historical Perspective." In *History of European Scientific and Technological Cooperation*, edited by J. Krige and L. Guzzetti, 323–345. Luxembourg: European Communities, 1997a.

Raphael, L. "Die Verwissenschaftlichung des Sozialen als methodische und konzeptionelle Herausforderung für eine Sozialgeschichte des 20. Jahrhunderts." *Geschichte und Gesellschaft* 22, no. 2 (1996): 165–193.

Riordan, M., Hoddeson, L., and Kolb, A. *Tunnel Visions: The Rise and Fall of the Superconducting Super Collider.* Chicago: University of Chicago Press, 2015b.

Ritter, G. *Großforschung und Staat in Deutschland: Ein historischer Überblick.* München: Beck, 1992b.

Roland, J.-L. "COST: An Unexpected Successful Cooperation." In *History of European Scientific and Technological Cooperation*, eds. J. Krige and L. Guzzetti, 355–368. Luxembourg: European Communities, 1997.

Schipper, F., and Schot, J. "Infrastructural Europeanism, or the Project of Building Europe on Infrastructures: An Introduction." *History and Technology* 27, no. 3 (2011b): 245–264.

Servan-Schreiber, J.-J. *Le Défi Américain.* Paris: Denoël, 1967).

Stahlecker, T. and Kroll, H. *Policies to Build Research Infrastructures in Europe: Following Traditions or Building New Momentum?* Karlsruhe: Fraunhofer ISI, 2013.

Tassinari, F. *Variable Geometries. Mapping Ideas, Institutions and Power in the Wider Europe*, CEPS Working Document No. 254 (2006).

Technopolis Group. *Big Science and Innovation*. Brighton: Technopolis Group, 2013.

Tindemans, P. "Post-War Research, Education and Innovation Policy-Making in Europe." In *European Science and Technology Policy: Towards Integration or Fragmentation?* edited by H. Delanghe, U. Muldur, and L. Soete, 9–18. Cheltenham: Edward Elgar, 2009.

Trischler, H. "Physics and Politics. Research and Research Support in Twentieth Century Germany in International Perspective – An Introduction." In *Physics and Politics: Research and Research Support in Twentieth Century Germany in International Perspective*, edited by H. Trischler and M. Walker, 9–18. Stuttgart: Franz Steiner Verlag, 2010.

Trischler, H., and Weinberger, H. "Engineering Europe: Big Technologies and Military Systems in the Making of 20th Century Europe." *History and Technology* 21, no. 1 (2005b): 49–83.

Ulnicane, I. "Research and Innovation as Sources of Renewed Growth? EU Policy Responses to the Crisis." *Journal of European Integration* 38, no. 3 (2016): 327–341.

Ulnicane, I. "Broadening Aims and Building Support in Science, Technology and Innovation Policy: The Case of the European Research Area." *Journal of Contemporary European Research* 11, no. 1 (2015): 31–49.

van der Vleuten, E. "Large Technical Systems." In *A Companion to the Philosophy of Technology*, edited by J. Friis, S. Pedersen, and V. Hendricks, 218–222. Chichester: Wiley-Blackwell, 2009.

van der Vleuten, E. "Toward a Transnational History of Technology: Meanings, Promises, Pitfalls." *Technology and Culture* 49, no. 4 (2008): 974–994.

Weingart, P. "Scientific Expertise and Political Accountability: Paradoxes of Science in Politics." *Science and Public Policy* 26, no. 3 (1999): 151–161.

Weingart, P. "Verwissenschaftlichung der Gesellschaft – Politisierung der Wissenschaft." *Zeitschrift für Soziologie* 12 (1983): 225–241.

Widmalm, S. "Big Science in a Small Country: Sweden and CERN II." In *Center on the Periphery: Historical Aspects of 20th-Century Swedish Physics*, edited by S. Lindqvist, 107–140. Canton: Warson Publishing International, 1993.

Williams, R. "Choosing Big Technologies: The Core Issues." In *Choosing Big Technologies*, edited by J. Krige, 223–234. Chur: Harwood Academic Publishers, 1993.

Chapter 3

Adriaens, A. "Non-Destructive Analysis and Testing of Museum Objects: An Overview of 5 Years of Research." Spectrochimica Acta Part B 60, no. 12 (2005): 1503–1516.

Altarelli, M. "Physics with Third Generation Synchrotron Sources." Physica Scripta T55 (1994): 9–13.

Andruszkow, J., Aune, B., Ayvazyan, V., Baboi, N., Bakker, R., Balakin, V., Barni, D., Bazhan, A., Bernard, M., Bosotti, A., et al. "First Observation of Self-Amplified Spontaneous Emission in a Free-Electron Laser at 109 nm Wavelength." Physical Review Letters 85, no. 18 (2000a): 3825–3829.

Balewski, K., Brefeld, W., Decking, W., Franz, H., Röhlsberger, R. and Weckert, E., eds. *PETRA III: A Low Emittance Synchrotron Radiation Source: Technical Design Report* (Executive Summary). Hamburg: DESY, 2004.

Belakhovsky, M. "Histoire et Développement des Sources de Rayonnement Synchrotron." Reflets de la Physique, 34–35 (2013): 10–11.

Billardon, M., Elleaume, P., Ortega, J., Bazin, C., Bergher, M., Velghe, M., Petroff, Y., Deacon, D., Robinson, K., Madey, J. "First Operation of a Storage-Ring Free-Electron Laser." Physical Review Letters 51, no. 18 (1983): 1652–1655.

Blaustein, R. "Biology and Light Sources: Synchrotrons Allow Researchers a Deep Look into Life." BioScience 67, no. 3 (2017): 201–207.

Bonifacio, R., Pellegrini, C., and Narducci, L. "Collective Instabilities and High-Gain Regime in a Free Electron Laser." Optics Communications 50, no. 6 (1984): 373–378.

Cramer, K. C. "Lightening Europe: Establishing the European Synchrotron Radiation Facility (ESRF)." History and Technology 33, no. 4 (2017b): 396–427.

Dhez, P. "Synchrotron Radiation in France: The Early Years." Synchrotron Radiation News 28, no. 4 (2015): 42–43.

Drobyazko, I., Kulipanov, G., Litvinenko, V., Pinayev, I., Popik, V., Silvestrov, I., Skrinsky, A., Sokolov, A., and Vinokurov, N. "Lasing in Visible and Ultraviolet Regions in an Optical Klystron Installed on the VEPP-3 Storage Ring." Nuclear Instruments and Methods in Physics Research Section A 282, 2–3 (1989): 424–430.

Eberhardt, W. "Synchrotron Radiation: A Continuing Revolution in X-Ray Science – Diffraction Limited Storage Rings and Beyond." Journal of Electron Spectroscopy and Related Phenomena 200 (2015): 31–39.

European XFEL, Enlightening Science (Schenefeld: European XFEL GmbH, 2017).

Feldhaus, J., Arthur, J., and Hastings, J. "X-Ray Free-Electron Lasers." Journal of Physics B 38, no. 9 (2005): 799–819.

Fox, M. Quantum Optics: An Introduction. Oxford, New York: Oxford University Press, 2006.

Freund, H. and Antonsen, T. Principles of Free-Electron Lasers. Dordrecht: Springer Netherlands, 1992.

Giannessi, L. and Masciovecchio, C. "FERMI: Present and Future Challenges." Applied Sciences 7, no. 6 (2017): 640.

Hallonsten, O. "Growing Big Science in a Small Country: MAX-Lab and the Swedish Research Policy System." Historical Studies in the Natural Sciences 41, no. 2 (2011): 179–215

Hallonsten, O. "The Parasites." Historical Studies in the Natural Sciences 45, no. 2 (2015b): 217–272.

Hallonsten, O. Big Science Transformed: Science, Politics and Organization in Europe and the United States. Cham: Palgrave Macmillan, 2016c.

Hallonsten, O. and Heinze, T. "Formation and Expansion of a New Organizational Field in Experimental Science." Science and Public Policy 42, no. 6 (2015): 841–854.

Hallonsten, O. and Heinze, T. "From Particle Physics to Photon Science: Multidimensional and Multi-Level Renewal at DESY and SLAC." Science and Public Policy 40, no. 5 (2013a): 591–603.

Heinze, T., Hallonsten, O., and Heinecke, S. "From Periphery to Center: Synchrotron Radiation at DESY, Part I: 1962–1977." Historical Studies in the Natural Sciences 45, no. 3 (2015a): 447–492.

Heinze, T., Hallonsten, O., and Heinecke, S. "From Periphery to Center.: Synchrotron Radiation at DESY, Part II: 1977–1993." *Historical Studies in the Natural Sciences* 45, no. 4 (2015c): 513–548.

Hoddeson, L., Kolb, A., and Westfall, C. Fermilab: Physics, the Frontier, and Megascience. Chicago: University of Chicago Press, 2008c.

Huang, Z. and Schmüser, P. "Free-Electron Lasers." In Handbook of Accelerator Physics and Engineering, edited by A. W. Chao, K. H. Mess, T. Maury, and F. Zimmermann, 229–233. Singapore: World Scientific, 2013.

Komitee für Forschung mit Synchrotronstrahlung. *Forschung mit Synchrotronstrahlung in Deutschland: Status und Perspektiven.* 2009. Online available: https://www.sni-portal.de/de/Dateien/forschung-mit-synchrotronstrahlung-in-deutschland-status-und-perspektiven-2/at_download/file, last accessed 1 March 2020.

Kondratenko, A. and Saldin, E. "Generation of Coherent Radiation by a Relativistic Electron Beam in an Ondulator." Particle Accelerators 10 (1980): 207–216.

Kunz, C. "Introduction: Properties of Synchrotron Radiation." In *Synchrotron Radiation: Techniques and Applications*, edited by C. Kunz, 1–23. Berlin, Heidelberg: Springer Verlag 1979.

Kunz, C. Synchrotronstrahlung bei DESY: Anfänge. Private print, 2012a.

Kunz, C., Skibowski, M., and Sonntag, B. "How It All Started at DESY in 1964." Synchrotron Radiation News 28, no. 4 (2015): 16–19.

Liénard, A.-M. "Champ Électrique et Magnétique Produit par une Charge Électrique Concentrée en un Point et Animée d'un Mouvement Quelconque" *L'Éclairage Électrique* 16, no. 27 (1898): 5–14.

Liss, K.-D. "Facilities in Quantum Beam Science." *Quantum Beam Science* 2, *no.* 6 (2018): 1–4.

Liss, K.-D. "Quantum Beam Science – Applications to Probe or Influence Matter and Materials." Quantum Beam Science 1, no. 1 (2017): 1–5.

Liss, K.-D. and Chen, K. "Frontiers of Synchrotron Research in Materials Science." MRS Bulletin 41, no. 6 (2016): 435–441.

Lohrmann, E. and Söding, P. Von schnellen Teilchen und hellem Licht: 50 Jahre Deutsches Elektronen-Synchrotron DESY. Weinheim: Wiley, 2009b.

Lynch, D., Plummer, W., Himpsel, F., Chiang, T., Margaritondo, G., and Lapeyre, G. "Tantalus, the First Dedicated Synchrotron Radiation Source." Synchrotron Radiation News 28, no. 4 (2015): 20–23.

Materlik, G. and Tschentscher, T., eds. TESLA Technical Design Report: Part V: The XFEL. Hamburg: DESY, 2001a.

Michalowicz, A., Ostrowiecki, H. and Martelly, I. "Bref Historique sur les Instruments." Histoire de la Recherche Contemporaine 3, no. 1 (2014): 33–36.

Munro, I. "Fifty Years of Synchrotron Radiation Research in the UK." *Philosophical Transactions of the Royal Society* 377, no. 2147 (2019). Online available: https://royalsocietypublishing.org/doi/pdf/10.1098/rsta.2018.0230, last accessed 20 March 2020.

Pellegrini, C. "The History of X-Ray Free-Electron Lasers." *The European Physical Journal* H 37, no. 5 (2012a): 659–708.

Pellegrini, C. "X-Ray Free-Electron Lasers: From Dream to Reality." Physica Scripta T169 (2016): 1–22.

Pollock, H. "The Discovery of Synchrotron Radiation." American Journal of Physics 51, no. 3 (1983b): 278–280.

Roychoudhuri, C., Kracklauer, A. F., and Creath K. The Nature of Light: What is a Photon? Boca Raton: CRC Press, 2008.

Schoenlein, R., Boutet, S., Minitti, M., and Dunne, A. "The Linac Coherent Light Source: Recent Developments and Future Plans." Applied Sciences 7, no. 8 (2017): 850.

Vylet, V. and Liu, J. *Synchrotron Facilities and Free Electron Lasers*, SLAC-PUB-13049. Stanford: SLAC, 2007.

Winick, H. "Properties of Synchrotron Radiation." In Synchrotron Radiation Research, edited by H. Winick and S. Doniach, 11–27. Boston, MA: Springer, 1980.

Winick, H. *Fourth Generation Light Sources.* Paper presented at the 17th IEEE Particle Accelerator Conference (PAC 97); Vancouver, Canada, May 12–16, 1997.

Winick, H. *Synchrotron Radiation Sources: Present Capabilities and Future Directions*, SLAC-PUB-777 1. Stanford: SLAC, 1998a.

Winick, H., Doniach, S. "An Overview of Synchrotron Radiation Research." In Synchrotron Radiation Research, edited by H. Winick and S. Doniach, 1–10. Boston, MA: Springer, 1980b.
Yabashi, M., Tanaka, H., Tono, K., and Ishikawa, T. "Status of the SACLA Facility." Applied Sciences 7, no. 6 (2017): 604.
Zhentang, Z., Dong, W., Qiang, G., Lixin, Y., Ming, G., Yongbin, L., and Bo, L. "Status of the SXFEL Facility." Applied Sciences 7, no. 6 (2017): 607

Chapter 4

ESRF Archive, Grenoble, France

European Science Foundation. *Synchrotron Radiation: A Perspective View for Europe*. Strasbourg: European Science Foundation, 1977.
European Science Foundation. *European Synchrotron Radiation Facility: The Feasibility Study*. Strasbourg: European Science Foundation, 1979.
J. Als-Nielsen et al., The Case for a European Synchrotron Radiation Facility. 1982.
European Synchrotron Radiation Project: *European Synchrotron Radiation Facility. Report of the ESRP* (presented by B. Buras and S. Tazzari), October 1984.
Le Livre Blanc d'un Contrat Rompu. Le Laboratoire Européen de Rayonnement Synchrotron. Candidature de Strasbourg, 1984.
ESRF Foundation Phase Report. Grenoble: European Synchrotron Radiation Facility, February 1987.

Archives Nationales de France, Paris, France

Box No. 1990 0594/23: Direction Générale de la Recherche et de la Technologie
Box No. 19920550/3: Ministère ou Secrétariat d'État Chargé de la Recherche
Box No. 19940426/2: Ministère ou Secrétariat d'État Chargé de la Recherche
Box No. 19940426/10: Ministère ou Secrétariat d'État Chargé de la Recherche
Box No. 19980422/1: Ministère ou Secrétariat d'État Chargé de la Recherche
Box No. 19980422/3: Ministère ou Secrétariat d'État Chargé de la Recherche
Box No. 20000405/14: Direction Générale de la Recherche et de la Technologie
Box No. 20000405/15: Direction Générale de la Recherche et de la Technologie
Box No. 20111003/205-207: Centre National de la Recherche Scientifique

Bundesarchiv, Koblenz, Germany

B 196/ 146 158: European Synchrotron Radiation Facility (ESRF) 1977–1983
B 196/ 146 157: ESRF Standortfragen 28.10.83-30.4.88 Band 1
B 196/ 121882: Paris Mai 1983 - Treffen von Minister Riesenhuber mit Minister Fabius
B 196/ 121883: Bonn Nov. 1983 - Treffen von Minister Riesenhuber mit Minister Fabius
B 196/ 121884: Rambouillet Mai 1984 - Treffen von Minister Riesenhuber mit Minister Fabius
B 196/ 122372: Minister Riesenhuber und Minister Fabius im Okt. 1983 in Bonn

Legal/Governmental Documents

Agreement establishing the European Molecular Biology Laboratory, EMBL (13 February 1969).
Agreement Concerning the Accession of the Government of the United Kingdom of Great Britain and Northern Ireland to the Convention of 19 January 1967 as amended by the Protocol of 6 July 1971 between the Government of the French Republic and the Government of the Federal Republic of Germany on the Construction and Operation of a Very High Flux Reactor (signed 19 July 1974).
Ministre Chargé du Plan et de l'Aménagement du Territoire, *Decision of the French Conseil d'Etat from* 8 January 1988, Request No. 74361.
Convention Concerning the Construction and Operation of a European Synchrotron Radiation Facility, ESRF (signed 16 December 1988/ 9 Dyecember 1991)

Literature

Atkinson, H. "Commentary on the History of ILL and ESRF." In History of European Scientific and Technological Cooperation. Edited by J. Krige and L. Guzzetti, 144–153. Luxembourg: European Communities, 1997.
Bacon, G. "Introduction – The Pattern of 50 Years." In Fifty Years of Neutron Diffraction: The Advent of Neutron Scattering. Edited by G. Bacon, 1–10. Bristol: Adam Hilger, 1987.
Bitsch, M.-T., ed. Le Couple France-Allemagne et les Institutions Européennes: Une Postérité pour le Plan Schuman? Bruxelles: E. Bruylant, 2001.
Central Intelligence Agency. *Special Report. Western European Space Programs.* 1964. Available at CIA FOIA database: https://www.cia.gov/library/readingroom/document/cia-rdp7900927a004400100002-3, last accessed 4 February 2020.

Collins, P. The Royal Society and the Promotion of Science since 1960. Cambridge: Cambridge University Press, 2016.

Cramer, K. C. "Lightening Europe: Establishing the European Synchrotron Radiation Facility (ESRF)." History and Technology 33, no. 4 (2017c): 396–427.

Defrance, C. "France-Allemagne: Une Coopération Scientifique 'Privilégiée' en Europe, de l'Immédiat Après-Guerre au Milieu des Années 1980?" In La Guerre Froide et l'Internationalisation des Sciences: Acteurs, Réseaux et Institutions. Edited by C. Defrance and A. Kwaschik, 169–186. Paris: CNRS, 2016c.

Defrance, C. and Pfeil, U. "Le Traité de l'Élysée et les Relations Scientifiques Franco-Allemandes." *Histoire de la Recherche Contemporaine* 2, no. 2 (2013). Online available: https://journals.openedition.org/hrc/369?lang=en, last accessed 30 December 2019.

Dumas, R. *Le* Fil et la Pelote: Mémoires. Paris: Plon, 1996.

Encrenaz, P., Gómez-González, J., Lequeux, J., and Orchiston, W. "Highlighting the History of French Radio Astronomy 7: The Genesis of the Institute of Radioastronomy at Millimeter Wavelengths (IRAM)." Journal of Astronomical History and Heritage 14, 2 (2011): 83–92.

European Commission. Objectives and Instruments of a Common Policy for Scientific Research and Technological Development. Brussels: European Commission, 1972.

Farge, Y. "L'Élaboration du Projet ESRF." *Histoire de la Recherche Contemporaine* 1, no. 1 (2012). Online available: https://journals.openedition.org/hrc/185, last accessed 29 December 2019.

Green, J., Quest, J. "A Short History of the European Transonic Wind Tunnel (ETW)." Progress in Aerospace Sciences 47, no. 5 (2011): 319–368.

Greenberg, D. and Boffey, P. "Europeans Set Up Science Foundation." Change 5, no. 10 (1973/1974): 78–79.

Grewing, M. "Selecting and Scheduling Observations at the IRAM Observatories." In Organizations and Strategies in Astronomy. Edited by A. Heck, 203–225. Dordrecht, London: Kluwer Academic Publishers, 2007.

Gutachterausschuss "Synchrotronstrahlung" bei der Koordinierungsstelle Hochenergiephysik. *Speicherringe für Synchrotron Strahlung. Bericht einer Studiengruppe.* 1977.

Guzzetti, L. A Brief History of European Union Research Policy. Luxembourg: European Communities, 1995b.

Hallonsten, O. and Heinze, T. "From Particle Physics to Photon Science: Multidimensional and Multi-Level Renewal at DESY and SLAC." Science and Public Policy 40, no. 5 (2013b): 591–603.

Hallonsten, O. Big Science Transformed: Science, Politics and Organization in Europe and the United States. Cham: Palgrave Macmillan, 2016d.

Hallonsten, O. Small Science on Big Machines: Politics and Practices of Synchrotron Radiation Laboratories. Lund: Research Policy Institute, 2009b.

Hirschel, E.-H., Prem, H. and Madelung, G. Aeronautical Research in Germany: From Lilienthal until Today. Heidelberg: Springer, 2004.

Jacrot, B. Des Neutrons pour la Science: Histoire de l'Institut Laue-Langevin, une Coopération Internationale Particulièrement Réussie. Les Ulis: EDP Sciences, 2006.

Kaelble, H. Les Relations Franco-Allemandes de 1945 à Nos Jours: Défis, Acquis, Options Nouvelles. Ostfildern: Thorbecke, 2004.

Krige, J. "Survey of Developments." In History of CERN: Volume I – Launching the European Organization for Nuclear Research. Edited by A. Hermann, L. Belloni, U. Mersists, D. Pestre, and J. Krige, 209–236. Amsterdam: North Holland, 1987.

Krige, J. "The Birth of EMBO and the Difficult Road to EMBL." Studies in History and Philosophy of Biological and Biomedical Sciences 22 (2002): 547–564.

Krige, J. "The Ford Foundation, European Physics and the Cold War." Historical Studies in the Physical and Biological Sciences 29, *no.* 2 (1999): 333–361.

Krige, J. "The Politics of European Scientific Collaboration." In Companion to Science in the Twentieth Century. Edited by J. Krige and D. Pestre, 897–918. London: Routledge, 2003c.

Krotz, U. and Schild, J. Shaping Europe: France, Germany, and Embedded Bilateralism from the Elysée Treaty to Twenty-First Century Politics. Oxford: Oxford University Press, 2013b.

Lappenküper, U. "On the Path to a "Hereditary Friendship"? Franco-German Relations Since the End of the Second World War." In History of Franco-German Relations in Europe: From "Hereditary Enemies" to Partners. Edited by C. Germond and H. Türk, 151–164. New York: Palgrave Macmillan, 2008.

Lellouche, P. "France-Allemagne: Le Double Déni." *Politique Étrangère*, no. 4 (2012): 739–754.

Middlemas, K. Orchestrating Europe: The Informal Politics of the European Union 1973–1995. London: Fontana Press, 1995b.

Papon, P. "European Scientific Cooperation and Research Infrastructures: Past Tendencies and Future Prospects." Minerva 42, no. 1 (2004c): 61–76.

Papon, P. "Intergovernmental Cooperation in the Making of European Research." In European Science and Technology Policy: Towards Integration or Fragmentation? Edited by H. Delanghe, U. Muldur, and L. Soete, 24–43. Cheltenham: Edward Elgar, 2009c.

Papon, P. "L'Espace Européen de la Recherche (1960–1985): Entre Science et Politique." In La Construction d'un Espace Scientifique Commun? La France, la RFA et l'Europe après le "Choc du Spoutnik". Edited by C. Defrance and U. Pfeil, 37–54. Bruxelles, New York: P.I.E. Peter Lang, 2012d.

Papon, P. "La Politique Européenne du CNRS dans les Années 1980." *Histoire de la Recherche Contemporaine*, Tome I-N°1 (2012e). Online available: https://journals.openedition.org/hrc/183, last accessed 10 March 2020.
Papon, P. L'Europe de la Science et de la Technologie. Grenoble: Press Universitaires de Grenoble (PUG), 2001b.
Pestre, D. "The Prehistory of the Franco-German Laue-Langevin Institute." In History of European Scientific and Technological Cooperation. Edited by J. Krige and L. Guzzetti, 137–143. Luxembourg: European Communities, 1997.
Peterson, J. "Eureka: A Historical Perspective." In History of European Scientific and Technological Cooperation. Edited by J. Krige and L. Guzzetti, 323–345. Luxembourg: European Communities, 1997b.
Rembser, J. "Neue Großgeräte für die Grundlagenforschung in der Bundesrepublik Deutschland." Physikalische Blätter 40, no. 5 (1984): 115–120.
Saunier, G. "La Genèse du Premier Programme-Cadre Européen: Un Regard Francais (1981–1984)." In La Construction d'un Espace Scientifique Commun? La France, la RFA et l'Europe après le "Choc du Spoutnik". Edited by C. Defrance and U. Pfeil, 81\–95. Bruxelles, New York: P.I.E. Peter Lang, 2012
Saunier, G. "Prélude à la Relance de l'Europe. Le Couple Franco-Allemand et les Projets de Relance Communautaire Vus de l'Hexagone (1981–1985)." In Le Couple France-Allemagne et les Institutions Européennes: Une Postérité pour le Plan Schuman? Edited by M.-T. Bitsch, 463–485. Bruxelles: E. Bruylant, 2001.
Schmied, H. "The European Synchrotron Radiation Story – Phase II." Synchrotron Radiation News 3, no. 6 (1990): 22–26.
Strasser, B. "The Coproduction of Neutral Science and Neutral State in Cold War Europe: Switzerland and International Scientific Cooperation, 1951–69,". Osiris 24 (2009): 165–187.
Strasser, B. "The Transformation of the Biological Sciences in Post-War Europe." EMBO Reports 4, no. 6 (2003): 540–543.
Taylor, P. The European Union in the 1990s. Oxford: Oxford University Press, 1996.
Trischler, H., Weinberger, H. "Engineering Europe: Big Technologies and Military Systems in the Making of 20th Century Europe." History and Technology 21, no. 1 (2005c): 49–83.
van der Bliek, J. ETW, a European Resource for the World of Aeronautics: The History of ETW in the Context of European Aeronautical Research and Development. Cologne: ETW, 1996.
Weidenfeld, W. and Wessels, W., eds. Jahrbuch der Europäischen Integration 1985. Bonn: Europa Union Verlag, 1986.
Witte, K. *History of the ESRF*, manuscript of a talk given at the Science Days in Aussois, May 2001.

Chapter 5

Bundesarchiv, Koblenz, Germany

B 304/1731 2: Akten des Sekretariats der Ständigen Konferenz der Kultusminister der Länder in der Bundesrepublik Deutschland.

DESY Archive, Hamburg, Germany

DESY. *Planungsschritte Linearcollider.*

TESLA Collaboration. *TTF Collaboration Meeting – Final Report*, June 1–2 1993, Hamburg.

Niederschrift über die 78. Sitzung des Verwaltungsrates am 13. Juli 1993 in Zeuthen.

Niederschrift über die 82. Sitzung des Verwaltungsrates am 28. Juni 1995 in Zeuthen.

Niederschrift über die 87. Sitzung des Verwaltungsrates am 12. November 1997 in Zeuthen.

Niederschrift über die 95. Sitzung des Verwaltungsrates am 5. Dezember 2001 in Hamburg.

DESY. *Memorandum on the Realisation of the First Stage of the European TESLA XFEL LABORATORY*, 24 October 2002.

Niederschrift über die 98. Sitzung des Verwaltungsrates am 12./13. Juni 2003 in Zeuthen.

100. Sitzung des DESY Verwaltungsrates am 24./25. Juni 2004 bei DESY in Zeuthen.

Bericht des Direktoriums zur 102. Sitzung des DESY Verwaltungsrates am 2./3. Dezember 2004 bei DESY in Hamburg.

Legal/Governmental Documents

Bürgerschaft der Freien und Hansestadt Hamburg, *Mitteilung des Senats an die Bürgerschaft. Gesetz zum Staatsvertrag zwischen der Freien und Hansestadt Hamburg und dem Land Schleswig-Holstein über die Schaffung der planerischen Voraussetzungen für die Errichtung und den Betrieb eines Linearbeschleunigers*, Drucksache 16/436, 4 November 1998.

Bundestag, *Antrag der Abgeordneten Ulrike Flach et al. und der Fraktion der F.D.P.: Forschungsstandort Deutschland stärken – Zukunftsprojekt „TESLA" nicht gefährden*, Drucksache 14/4646, 15 November 2000.

Memorandum of Understanding Between the Deutsches Elektronen-Synchrotron (DESY) and the Stanford Linear Accelerator Center (SLAC) Establishing a Collaborative Research Effort to Enable the Exploitation and Expansion of the

Scientific Capabilities of The Linac Coherent Light Source and The TESLA X-Ray Free-Electron Laser, 1 November 2002.
Staatsvertrag zwischen der Freien und Hansestadt Hamburg und dem Land Schleswig-Holstein über die Schaffung der planerischen Voraussetzungen für die Errichtung und den Betrieb eines Freie Elektronen Lasers im Röntgenlaserbereich, 28 September 2004.
Gemeinsame Erklärung über die strategische Partnerschaft auf dem Gebiet der Bildung, Forschung und Innovation zwischen der Bundesrepublik Deutschland und der Russischen Föderation, 11 April 2005.
Agreement on the Establishment of the ITER International Fusion Energy Organization for the Joint Implementation of the ITER Project, 21 November 2006.
Articles of Association of the "European X-Ray Free-Electron Laser Facility GmbH" (European XFEL GmbH), 30 November 2009.
Convention Concerning the Construction and Operation of a European X-Ray Free-Electron Laser Facility, 30 November 2009.
Final Act of the Conference of Plenipotentiaries for the Establishment of a European X-Ray Free-Electron Laser Facility, 30 November 2009.
Convention concerning the Construction and Operation of a Facility for Antiproton and Ion Research in Europe (FAIR), 12 October 2010.
Schleswig-Holsteinischer Landtag, *Gesetzentwurf der Landesregierung zu dem Abkommen zwischen der Bundesrepublik Deutschland, der Freien und Hansestadt Hamburg und dem Land Schleswig-Holstein zum Bau und Betrieb der Europäischen Freie-Elektronen-Röntgenlaseranlage (XFEL) und dem Staatsvertrag zwischen der Freien und Hansestadt Hamburg und dem Land Schleswig-Holstein über den Bau und Betrieb der Europäischen Freie-Elektronen-Röntgenlaseranlage (XFEL)*, 15 January 2010.
Letter of Intent of the Deutsches Elektronen-Synchrotron DESY (Hamburg, Germany) and the National Research Centre "Kurchatov Institute" (Moscow, Russian Federation) Concerning the German-Russian Cooperation in Strategic Research Fields: The Ioffe-Röntgen Institute, 23 May 2011.
Protocole d'Adhésion du Gouvernment de la Féderation de Russie à la Convention du 16 Décembre 1988 Relative à la Construction et à l'Exploitation d'une Installation Européenee de Rayonnement de Synchrotron, 23 June 2014.
Satzung des Vereins "Hermann von Helmholtz-Gemeinschaft Deutscher Forschungszentren e.V." (Von der Gründungsversammlung am 12. September 2001 festgestellte Fassung mit den von der Mitgliederversammlung auf ihrer 79. Sitzung am 15. Oktober 2003, auf ihrer 87. Sitzung am 13. September 2006, auf ihrer 99. Sitzung am 19. September 2012 und auf ihrer 111. Sitzung am 11. September 2018 beschlossenen Änderungen, eingetragen in das Vereinsregister unter VR 7942 beim Amtsgericht Bonn am 17. Oktober 2018), 11 September 2018.

References

Adomeit, H. "German-Russian Relations: Balance Sheet since 2000 and Perspectives until 2025." *ifri. Comité d'Études des Relations Franco-Allemandes*, 2012.

Andruszkow, J., Aune, B., Ayvazyan, V., Baboi, N., Bakker, R., Balakin, V., Barni, D., et al. "First Observation of Self-Amplified Spontaneous Emission in a Free-Electron Laser at 109 nm Wavelength." *Physical Review Letters* 85, no. 18 (2000b): 3825–3829.

Banks, M. "Funding Shortfall Hits European X-ray Laser Facility." *Physics World* (29 July 2011). Online available: http://physicsworld.com/cws/article/news/2011/jul/29/funding-shortfall-hits-european-x-ray-laser-facility, last accessed 20 May 2019.

Bartz, O. *Der Wissenschaftsrat: Entwicklungslinien der Wissenschaftspolitik in der Bundesrepublik Deutschland 1957–2007* (Stuttgart: Franz Steiner Verlag, 2007).

Behnke, T., Brau, J., Foster, B., Fuster, B., Harrison, M., McEwan Paterson, J., Peskin, M., Stanitzki, M., Walker, N., and Yamamoto, H. (eds.) *The International Linear Collider: Technical Design Report.* 2013.

Bonucci, A. *In-Kind Contributions to the European XFEL Facility*, presentation given at the 2nd In-Kind Contributions Workshop (IKCW 2015) at FAIR. Online available: https://indico.gsi.de/event/3426/, last accessed 1 May 2019.

Brinkmann, R., Materlik, G., Rossbach, J., and Wagner, A. (eds.) *Conceptual Design Report of a 500 GeV e^+e^- Linear Collider with Integrated X-Ray Laser Facility.* Hamburg: DESY, 1997.

Brinkmann, R., Schmüser, P., and Wiik, B. "TESLA." *Physik in unserer Zeit* 30, no. 2 (1999): 62–68.

Castells, M. and Cardoso, G. (eds.) *The Network Society: From Knowledge to Policy.* Washington DC: Johns Hopkins Center for Transatlantic Relations, 2005b.

CREMLIN. *First CREMLIN Recommendations for the European-Russian Megascience Collaboration.* Hamburg: DESY, 5 June 2018.

DESY XFEL Project Group. *The European X-Ray Free Electron Laser. Technical Design Report.* Hamburg: DESY XFEL Project Group/ European XFEL Project Team, 2007.

DESY. *Particle Physics 2016: Jahresbericht des Bereichs Teilchenphysik bei DESY.* Online available: http://www.desy.de/sites2009/site_www-desy/content/e410/e84441/e234837/ParticlePhysics2016_ger.pdf, last accessed 10 November 2017.

DESY. *Wissenschaftlicher Jahresbericht 2004.* Hamburg: DESY, 2005.

DESY. *Worldwide Consensus of Particle Physicists: Superconductivity for Future Particle Accelerator Project ILC*, News Release, 20 August 2004. Online available: https://www.desy.de/information__services/press/press_releases/2004/pr_200804/index_eng.html, last accessed 20 March 2020.

DESY. *Wissenschaftlicher Jahresbericht 2003*. Hamburg: DESY, 2004.
DESY. *Wissenschaftlicher Jahresbericht 1998*. Hamburg: DESY, 1999.
DESY. *Wissenschaftlicher Jahresbericht 1995*. Hamburg: DESY, 1996.
DESY. *A VUV Free Electron Laser at the TESLA Test Facility at DESY: Conceptual Design Report*. Hamburg: DESY, 1995.
Elzinga, A. "Features of the Current Science Policy Regime: Viewed in Historical Perspective." *Science and Public Policy* 39, no. 4 (2012b): 416–428.
European Commission. *Roadmap for EU-Russia S&T Cooperation*. Brussels: European Commission, 2017.
European Commission. *Review of the S&T Cooperation Agreement between the European Union and Russia*. Brussels: European Commission, 2013.
European Committee for Future Accelerators. *Report 1967*. Geneva: CERN, 1967.
European XFEL. *Annual Report 2016*. Schenefeld: European XFEL GmbH, 2017.
European XFEL. *The United Kingdom Officially Joins European XFEL*, News Release, 19 March 2018. Online available: https://www.xfel.eu/news_and_events/news/index_eng.html?openDirectAnchor=1435&two_columns=0, last accessed 25 March 2018.
European XFEL. *UK Contributes a High-Energy Optical Laser to HED Instrument*, News Release 21 July 2015. Online available: https://www.xfel.eu/news_and_events/news/index_eng.html?openDirectAnchor=1182&two_columns=0, last accessed 10 March 2018.
European XFEL. *United Kingdom Withdraws Participation in European XFEL*, News Release 18 December 2009. Online available: https://www.xfel.eu/news_and_events/archived_news/2009/20091218/index_eng.html, last accessed 25 April 2017.
Forsberg, T. "From Ostpolitik to Frostpolitik? Merkel, Putin and German Foreign Policy towards Russia." *International Affairs* 92, no. 1 (2016): 21–42.
Galison, P. and Hevly B. (eds) *Big Science: The Growth of Large-Scale Research* (Stanford: Stanford University Press, 1992b).
Gorzka, G. *Knowledge Transfer between Germany and Russia: Drivers, Instruments and Impact*. Kassel: Kassel University Press, 2016.
Götz, R. "Deutschland und Russland – ‚strategische Partner'?" *APuZ (Aus Politik und Zeitgeschichte)* (March 2006): 14–23.
Graham, L. and Dezhina, I. *Science in the New Russia: Crisis, Aid, Reform*. Bloomington: Indiana University Press, 2008.
Greenberg, D. *The Politics of Pure Science*, new edition. Chicago: University of Chicago Press, 1999.
Gutbrod, H., Augustin, I., Eickhoff, H., Groß, K.-D., Henning, W., Krämer, D., and Walter G. (eds.) *FAIR Baseline Technical Report: Executive Summary*. Darmstadt: Gesellschaft für Schwerionenphysik, 2006.
Habfast, C. *Grossforschung mit kleinen Teilchen: Das Deutsche Elektronen-Synchrotron, DESY, 1956–1970*. Berlin, New York: Springer, 1989.

Hallonsten, O. *Big Science Transformed: Science, Politics and Organization in Europe and the United States.* Cham: Palgrave Macmillan, 2016e.

Hallonsten, O. "The Politics of European Collaboration in Big Science." In *The Global Politics of Science and Technology – Vol. 2.* Edited by M. Mayer, M. Carpes and R. Knoblich, 31–46. Berlin, Heidelberg: Springer, 2014c.

Hallonsten, O. "Continuity and Change in the Politics of European Scientific Collaboration." *Journal of Contemporary European Research* 8, no. 3 (2012b): 300–319.

Hallonsten, O. and Heinze, T. "Institutional Persistence Through Gradual Organizational Adaptation: Analysis of National Laboratories in the USA and Germany." *Science and Public Policy* 39, no. 4 (2012): 450–463.

Harringa, C. "Zwischen Völkerrecht und Frascati: Praktische Aspekte der Ausgestaltung internationaler Kooperationen am Deutschen Elektronen Synchrotron (DESY)." *Ordnung der Wissenschaft* 2 (2018): 69–74.

Heinecke, S. *Die deutsche Russlandpolitik 1991–2005: Entwicklungen und Gesellschaftliche Einflüsse in außenpolitischen Entscheidungsprozessen.* Doctoral Thesis, University of Cologne, 2011.

Heinze, T., and Arnold, N. "Governanceregimes im Wandel." *KZfSS Kölner Zeitschrift für Soziologie und Sozialpsychologie* 60, no. 4 (2008): 686–722.

Heinze, T., Hallonsten, O., and Heinecke, S. "Turning the Ship: The Transformation of DESY, 1993–2009." *Physics in Perspective* 19, no. 4 (2017b): 424–451.

Heinze, T., Hallonsten, O., and Heinecke, S. "From Periphery to Center: Synchrotron Radiation at DESY, Part II: 1977–1993." *Historical Studies in the Natural Sciences* 45, no. 4 (2015b): 513–548.

Heinze, T., Hallonsten, O., and Heinecke, S. "From Periphery to Center: Synchrotron Radiation at DESY, Part I: 1962–1977." *Historical Studies in the Natural Sciences* 45, no. 3 (2015d): 447–492.

Helling-Moegen, S. *Forschen nach Programm: Die programmorientierte Förderung in der Helmholtz-Gemeinschaft: Anatomie einer Reform.* Marburg: Tectum, 2009.

Hoddeson, L. "The First Large-Scale Application of Superconductivity: The Fermilab Energy Doubler, 1972–1983." Historical Studies in the Physical and Biological Sciences 18, no. 1 (1987): 25–54.

Hoddeson, L., Brown, L. M., Dresden, M., and Riordan, M. (eds.) *The Rise of the Standard Model: A History of Particle Physics from 1964 to 1979.* Cambridge: Cambridge University Press, 1997;

Hoffmann, D. and Trischler, H. "Die Helmholtz-Gemeinschaft in Historischer Perspektive." In *20 Jahre Helmholtz-Gemeinschaft.* Edited by J. Mlynek and A. Bittner, 9–47. Bonn: Helmholtz Gemeinschaft, 2015.

House of Commons/ Science and Technology Committee. *The Impact of Spending Cuts on Science and Scientific Research* (Sixth Report of Session 2009–10). London: House of Commons, 2010.

Kallerud, E., Klitkou, A., Sutherland Olsen, D., Scordato, L., Amanatidou, E., Upham, P., Nieminen, M., Lima-Toivanen, M. and Oksanen, J. *Dimensions of Research and Innovation Policies to Address Grand and Global Challenges.* NIFU Working Paper 13, 2013.

Karaulova, M., Gök, A., Shackleton, O., and Shapira, P. "Science System Path-Dependencies and their Influences: Nanotechnology Research in Russia." *Scientometrics* 107 (2016): 645–670.

Kehrt, C. *Mit Molekülen spielen: Wissenschaftskulturen der Nanotechnologie zwischen Politik und Medien* (Bielefeld: transcript Verlag, 2016b).

Kessler, V. *Die sichtbare Hand des Staates: Eine rechtsökonomische Analyse russischer Staatskorporationen.* Berlin: BWV Berliner Wissenschafts-Verlag, 2014.

Kevles, D. "Good-Bye to the SSC: On the Life and Death of the Superconducting Super Collider." *Engineering & Science* 58, no. 2 (1995): 16–25.

Kölbel, M. "Das Bundesministerium für Bildung und Forschung (BMBF) als Wissenschaftspolitischer Akteur." In *Handbuch Wissenschaftspolitik.* Edited by S. Hornbostel, A. Knie and D. Simon, 533–548. Wiesbaden: Verlag für Sozialwissenschaften, 2010.

Krige, J. *American Hegemony and the Postwar Reconstruction of Science in Europe.* Cambridge, MA: MIT Press, 2006b.

Kunz, C. *Synchrotronstrahlung bei DESY: Anfänge.* Private print, 2012b.

Lengwiler, M. "Kontinuitäten und Umbrüche in der Deutschen Wissenschaftspolitik des 20. Jahrhunderts." In *Handbuch Wissenschaftspolitik.* Edited by S. Hornbostel, A. Knie and D. Simon, 6–18. Wiesbaden: Verlag für Sozialwissenschaften, 2010b.

Lohrmann, E. and Söding, P. *Von schnellen Teilchen und hellem Licht: 50 Jahre Deutsches Elektronen-Synchrotron DESY.* Weinheim: Wiley, 2009c.

Lohrmann, E. and Söding, P. "DESY Marks 50 Years of Accelerator Research." *CERN Courier*, 7 December 2009. Online available: https://cerncourier.com/a/desy-marks-50-years-of-accelerator-research/, last accessed 10 March 2020.

Maiani, L. and Bonolis, L. "The LHC Timeline: A Personal Recollection (1980–2012)." *The European Physical Journal H* 42, 4–5 (2017): 475–505.

Materlik, G. and Tschentscher, T. (eds.) *TESLA Technical Design Report: Part V: The XFEL.* Hamburg: DESY, 2001b.

McCray, W. "'Globalization with Hardware': ITER's Fusion of Technology, Policy, and Politics." *History and Technology* 26, no. 4 (2010b): 283–312.

Medical Research Council. *UK invests in World's Most Advanced Crystallography Facility*, News Release, 3 June 2014. Online available: https://mrc.ukri.org/

news/browse/uk-invests-in-worlds-most-advanced-crystallography-facility, last accessed 10 April 2019.

Medical Research Council. *UK invests in World's Most Advanced Crystallography Facility*, News Release, 3 June 2014. Online available: https://mrc.ukri.org/news/browse/uk-invests-in-worlds-most-advanced-crystallography-facility, last accessed 10 April 2019.

Meusel, E.-J. "Die Förderung der Großforschung durch das BMBF." In *Das Wissensministerium: Ein halbes Jahrhundert Forschungs- und Bildungspolitik in Deutschland*. Edited by P. Weingart and N. Taubert, 144–157. Weilerswist: Velbrück Wissenschaft, 2006.

Neal, H., Smith, T., and McCormick, J. *Beyond Sputnik: U.S. Science Policy in the Twenty-First Century*. Ann Arbor: University of Michigan Press, 2008.

Neidhardt, F. "Institution, Organisation, Interaktion: Funktionsbedingungen des Wissenschaftsrats."*Leviathan* 40, no. 2 (2012): 271–296.

Nuhn, H.-D. "From Storage Rings to Free Electron Lasers for Hard X-Rays." *Journal of Physics: Condensed Matter* 16, no. 33 (2004): 13–21.

Oreskes, N. and Krige, J. (eds.) *Science and Technology in the Global Cold War*. Cambridge, MA: MIT Press, 2014.

Pellegrini, C. "The History of X-Ray Free-Electron Lasers." *The European Physical Journal H* 37, no. 5 (2012b): 659–708.

Prat, S. *Management and Control of In-Kind Contributions. Case Study: European XFEL Facility*, presentation given at RAMIRI 2012 Learning Programme, Trieste, 2012.

Prime Minister of the Russian Federation Vladimir Putin, *Prime Minister Vladimir Putin holds a session of the Government Commission on High Technology and Innovation in Dubna*, 5 July 2011, Archive of the Official Site of the 2008–2012 Prime Minister of the Russian Federation Vladimir Putin. Online available: http://archive.premier.gov.ru/eng/events/news/15785/video.html, last accessed 10 March 2020.

Radosevic, S. "Patterns of Preservation, Restructuring and Survival: Science and Technology Policy in Russia in Post-Soviet Era." *Research Policy*, no. 3 (2003): 1105–1124.

Riordan, M. "A Bridge Too Far: The Demise of the Superconducting Super Collider." *Physics Today* 69, no. 10 (2016): 48–54.

Riordan, M., Hoddeson, L. and Kolb, A. *Tunnel Visions: The Rise and Fall of the Superconducting Super Collider*. Chicago: University of Chicago Press, 2015c.

Ritter, G. *Großforschung und Staat in Deutschland: Ein historischer Überblick*. München: Beck, 1992c.

RUSNANO. *RUSNANO Takes Part in Construction and Operation of the Most Powerful in the World X-Ray Free Electron Laser*, News Release, 16 April 2009. Online available: http://en.rusnano.com/press-centre/news/88397, last accessed 12 January 2020.

Sakwa, R. "Vladimir Putin and Russian Foreign Policy Towards the West: Towards a New Realism." In *Russia and Europe in the Twenty-First Century: An Uneasy Partnership*. Edited by J. Gower and G. Timmins, 1–22. London: Anthem Press, 2009.

Schiermeier, Q. "High Hopes for Russia's Nanotech Firms." *Nature* 461, no. 7267 (2009): 1036–1037.

Schiermeier, Q. "Russia Pins its Hopes on 'Nano'." *Nature* 448, no. 7151 (2007): 233.

Schneider, J. "FLASH – From Accelerator Test Facility to the First Single-Pass Soft X-Ray Free-Electron Laser." *Journal of Physics B* 43, no. 19 (2010).

Schreiber, S. *Status of the TTF FEL*, paper presented at the 11th Workshop on RF Superconductivity, 8–12 September 2003 at Travemünde/Lübeck, Germany. Online available: https://accelconf.web.cern.ch/accelconf/SRF2003/papers/weo03.pdf, last accessed 10 March 2020.

Schummer, J. *Nanotechnologie – Spiele mit Grenzen*. Frankfurt a.M.: Suhrkamp, 2009.

Schunck, H. "Bundesweite Förderung der Physik in Deutschland." In *Physik im 21. Jahrhundert: Essays zum Stand der Physik*. Edited by W. Martienssen and D. Röß, 367–403. Berlin: Springer, 2011.

Schunck, H. *Meine Jahre mit DESY*. Private Print, 2014. Online available: https://bib-pubdb1.desy.de/record/397999, last accessed 10 March 2020.

Science Board. *Minutes of the Science Board's Meeting*, 7–8 December 2009. Online available: http://www.stfc.ac.uk/files/news-from-science-board-december-2009, last accessed 1 November 2019.

Shapin, S. *The Scientific Life: A Moral History of a Late Modern Vocation*. Chicago: University of Chicago Press, 2008b.

Siemann, R. "Uses of Superconductivity in Particle Accelerators." In *Lepton – Hadron Scattering, Proceedings*, 19th SLAC Summer Institute on Particle Physics (SSI 91), edited by J. Hawthorne, 21–46 (Stanford: SLAC).

Smirnov, S. "Russia – CERN Cooperation: Current Status and Perspectives." *Journal of Physics: Conference Series*, 1406, no. 1 (2019).

Stehr, N. *Knowledge Societies*. London: SAGE, 1994b.

STFC. *UK Joins World Leading X-Ray Laser Facility*, News Release, 19 March 2018. Online available: https://stfc.ukri.org/news/uk-joins-world-leading-x-ray-laser-facility, last accessed 3 April 2019.

Stucke, A. "Staatliche Akteure der Wissenschaftspolitik." In *Handbuch Wissenschaftspolitik*. Edited by S. Hornbostel, A. Knie and D. Simon, 485–501. Wiesbaden: Verlag für Sozialwissenschaften, 2010.

TESLA Collaboration, *TESLA XFEL First Stage of the X-Ray Laser Laboratory: Technical Design Report Supplement*. Hamburg: DESY, 2002.

TESLA Collaboration. *TESLA. The Superconducting Electron-Positron Linear Collider with an Integrated X-Ray Laser Laboratory: Technical Design Report.* Hamburg: TESLA Collaboration, 2001.

TESLA Collaboration. *Transparencies from the TESLA Meeting*, March 1–3, 1999, Hamburg, DESY. Online available: https://flash.desy.de/sites2009/site_vuvfel/content/e403/e1644/e1418/e1420/infoboxContent2286/tesla1999-06.pdf, last accessed 10 March 2020.

TESLA Collaboration. *A Proposal to Construct and Test Prototype Superconducting RF Structures for Linear Colliders.* Hamburg: DESY, 1992.

TESLA Test Facility FEL Team. *SASE FEL at the TESLA Facility, Phase 2* (TESLA-FEL 2002-01). Hamburg: DESY, 2002.

Timmins, G. and Gower, J. "Introduction: Russia and Europe: What Kind of Partnership?" *In Russia and Europe in the Twenty-First Century: An Uneasy Partnership.* Edited by J. Gower and G. Timmins, xxi–xxvi. London: Anthem Press, 2009b.

Westerlund, F. *Russian Nanotechnology R&D. Thinking Big About Small Scale Science.* Stockholm: Swedish Defence Research Agency, 2011.

Winick, H. *Synchrotron Radiation Sources: Present Capabilities and Future Directions*, SLAC-PUB-777 1, 1998b.

Wissenschaftsrat. *Planungssicherheit für die Wissenschaft zur Entscheidung des Bundes über Großgeräte der Naturwissenschaftlichen Grundlagenforschung*, News Release, 5 February 2003. Online available: https://www.wissenschafts-rat.de/download/archiv/PM_0503.pdf, last accessed 10 February 2020.

Wissenschaftsrat. *Stellungnahme zu neun Großgeräten der naturwissenschaftlichen Grundlagenforschung und zur Weiterentwicklung der Investitionsplanung von Großgeräten / Statement on Nine Large-Scale Facilities for Basic Scientific Research and on the Development of Investment Planning for Large-Scale Facilities.* Köln: Wissenschaftsrat, 2003.

Wissenschaftsrat. *Systemevaluation der HGF: Stellungnahme des Wissenschaftsrates zur Hermann von Helmholtz-Gemeinschaft Deutscher Forschungszentren.* Berlin: Wissenschaftsrat, 2001.

Wissenschaftsrat. *Perspektiven für Wissenschaft und Forschung auf dem Weg zur deutschen Einheit: Zwölf Empfehlungen.* Berlin: Wissenschaftsrat, 1990.

Witte, K. *Comparison of Legal Difficulties/Advantages of ESRF and European XFEL*, presentation given at Meeting N° 2 of CTA Legal Working Group, Paris, 2012.

Womersley, J. *The STFC Science Programme 2010–15*, presentation given at the STFC Town Meeting 2009. Online available: https://www.stfc.ac.uk/files/stfc-town-meeting, last accessed 20 October 2019.

Yegorov, I. "Post-Soviet Science: Difficulties in the Transformation of the R&D Systems in Russia and Ukraine." *Research Policy* 38, no. 4 (2009): 600–609.

Yoder, J. "From Amity to Enmity: German-Russian Relations in the Post Cold War Period." *German Politics and Society* 33, no. 3 (2015): 49–69.

Zubkovskaya, A. and Michailova, S. "The Development of Russian Multinational Enterprises from the 1990s to the Present." *Organizations and Markets in Emerging Economies* 5, no. 2 (2014): 59–78.

MISCELLANEOUS

Final Report Summary – PRE-XFEL (Preparatory Activities for the Implementation of the European X-Ray Free-Electron Laser Facility), Grant agreement ID: 211604, 1 March 2013. Online available: https://cordis.europa.eu/project/id/211604, last accessed 10 March 2020.

Website of the CREMLIN Project (Connecting Russian and European Measures for Large-Scale Research Infrastructures): https://www.cremlin.eu, last accessed 30 March 2020.

CHAPTER 6

Amos, J. "XFEL: Brilliant X-Ray Laser Comes Online." *BBC News*, 1 September 2017. Online available: https://www.bbc.com/news/science-environment-41117442, last accessed 12 December 2019.

Hallonsten, O. "Research Infrastructures in Europe: The Hype and the Field." *European Review* (2020b).

Joint Institute for Nuclear Research. "Russia and CERN are Working Out a New Format of Cooperation," News Release, 14 March 2018. Online available: http://www.jinr.ru/posts/russia-and-cern-are-working-out-a-new-format-of-cooperation, last accessed 20 March 2020.

Krige, J. "The Politics of European Scientific Collaboration." In *Companion to Science in the Twentieth Century*, edited by J. Krige and D. Pestre, 897–918. London: Routledge, 2003d.

Misa T., and Schot, J. "Inventing Europe: Technology and the Hidden Integration of Europe." *History and Technology* 21, no. 1 (2005c): 1–19.

Papon, P. "L'Espace Européen de la Recherche (1960–1985): Entre Science et Politique." In *La Construction d'un Espace Scientifique Commun? La France, la RFA et l'Europe après le "Choc du Spoutnik"*, edited by C. Defrance and U. Pfeil, 37–54. Bruxelles, New York: P.I.E. Peter Lang, 2012c.

Patel, K. K. "Provincialising European Union: Co-Operation and Integration in Europe in a Historical Perspective." *Contemporary European History* 22, no. 4 (2013b): 649–673.

Schipper, F., and Schot, J. "Infrastructural Europeanism, or the Project of Building Europe on Infrastructures: An Introduction." *History and Technology* 27, no. 3 (2011c): 245–264.

Trischler, H., and Weinberger, H. "Engineering Europe: Big Technologies and Military Systems in the Making of 20th Century Europe." *History and Technology* 21, no. 1 (2005d): 49–83.

TASS. "New Agreement with CERN to be Signed in 2018 – Russia's Education Ministry." News Release, 10 March 2018. Online available: http://tass.com/science/993418, last accessed 12 Mai 2020.

Miscellaneous

Petition of the Hungarian Synchrotron Committee at change.org: *Stop Hungary's withdrawal from the European Synchrotron Radiation Facility.* Online available: https://www.change.org/p/government-of-hungary-hungarian-academy-of-sciences-stop-hungary-s-withdrawal-from-the-european-synchrotron-radiation-facility, last accessed 30 March 2020.

Index[1]

[1] Note: Page numbers followed by 'n' refer to notes.

K. C. Cramer, *A Political History of Big Science*, Palgrave Studies in the History of Science and Technology, https://doi.org/10.1007/978-3-030-50049-8

E

N

O

P

R